Representations and Techniques for 3D Object Recognition and Scene Interpretation

Synthesis Lectures on Artificial Intelligence and Machine Learning

Editors

Ronald J. Brachman, *Yahoo! Research*
William W. Cohen, *Carnegie Mellon University*
Thomas Dietterich, *Oregon State University*

Representations and Techniques for 3D Object Recognition and Scene Interpretation
Derek Hoiem and Silvio Savarese
2011

A Short Introduction to Preferences: Between Artificial Intelligence and Social Choice
Francesca Rossi, Kristen Brent Venable, and Toby Walsh
2011

Human Computation
Edith Law and Luis von Ahn
2011

Trading Agents
Michael P. Wellman
2011

Visual Object Recognition
Kristen Grauman and Bastian Leibe
2011

Learning with Support Vector Machines
Colin Campbell and Yiming Ying
2011

Algorithms for Reinforcement Learning
Csaba Szepesvári
2010

Representations and Techniques for 3D Object Recognition and Scene Interpretation

Derek Hoiem and Silvio Savarese

ISBN: 978-3-031-00429-2   paperback
ISBN: 978-3-031-01557-1   ebook

DOI 10.1007/978-3-031-01557-1

A Publication in the Springer series
SYNTHESIS LECTURES ON ARTIFICIAL INTELLIGENCE AND MACHINE LEARNING

Lecture #15
Series Editors: Ronald J. Brachman, *Yahoo! Research*
               William W. Cohen, *Carnegie Mellon University*
               Thomas Dietterich, *Oregon State University*
Series ISSN
Synthesis Lectures on Artificial Intelligence and Machine Learning
Print 1939-4608   Electronic 1939-4616

Representations and Techniques for 3D Object Recognition and Scene Interpretation

Derek Hoiem
University of Illinois at Urbana-Champaign

Silvio Savarese
University of Michigan

SYNTHESIS LECTURES ON ARTIFICIAL INTELLIGENCE AND MACHINE LEARNING #15

ABSTRACT

One of the grand challenges of artificial intelligence is to enable computers to interpret 3D scenes and objects from imagery. This book organizes and introduces major concepts in 3D scene and object representation and inference from still images, with a focus on recent efforts to fuse models of geometry and perspective with statistical machine learning.

The book is organized into three sections: (1) Interpretation of Physical Space; (2) Recognition of 3D Objects; and (3) Integrated 3D Scene Interpretation. The first discusses representations of spatial layout and techniques to interpret physical scenes from images. The second section introduces representations for 3D object categories that account for the intrinsically 3D nature of objects and provide robustness to change in viewpoints. The third section discusses strategies to unite inference of scene geometry and object pose and identity into a coherent scene interpretation. Each section broadly surveys important ideas from cognitive science and artificial intelligence research, organizes and discusses key concepts and techniques from recent work in computer vision, and describes a few sample approaches in detail. Newcomers to computer vision will benefit from introductions to basic concepts, such as single-view geometry and image classification, while experts and novices alike may find inspiration from the book's organization and discussion of the most recent ideas in 3D scene understanding and 3D object recognition.

Specific topics include: mathematics of perspective geometry; visual elements of the physical scene, structural 3D scene representations; techniques and features for image and region categorization; historical perspective, computational models, and datasets and machine learning techniques for 3D object recognition; inferences of geometrical attributes of objects, such as size and pose; and probabilistic and feature-passing approaches for contextual reasoning about 3D objects and scenes.

KEYWORDS

object recognition, scene understanding, computer vision, single-view geometry, 3D scene models, 3D object models, image categorization, context

Contents

Preface

3D scene understanding and object recognition are among the grandest challenges in computer vision. A wide variety of techniques and goals, such as structure from motion, optical flow, stereo, edge detection, and segmentation, could be viewed as subtasks within scene understanding and recognition. Many of these applicable methods are detailed in computer vision books (e.g., [72, 91, 208, 224]), and we do not aim to repeat these details. Instead, we focus on high-level representations for scenes and objects, particularly representations that acknowledge the 3D physical scene that underlies the image. This book aims to help the reader answer the following questions:

- How does the 2D image relate to the 3D scene, and how can we take advantage of that perspective relation?

- How can we model the physical scene space, and how can we estimate scene space from an image?

- How can we represent and recognize objects in ways that are robust to viewpoint change?

- How can we use knowledge of perspective and scene space to improve recognition, or vice versa?

WHY WE SHOULD INTERPRET IMAGES AS PHYSICAL SCENES

Objects live and interact in the physical space of the scene, not in the pixel space of the image. If we want computers to describe objects, predict their actions, and interact with the environment, then we need algorithms that interpret images in terms of the underlying 3D scene. In some cases, such as 3D reconstruction and navigation, we care about the 3D interpretation itself. In others, such as object recognition and scene interpretation, we may not be particularly concerned about the depth of objects and surfaces, but a 3D interpretation can provide a more powerful context or frame for reasoning about physical interaction. Here, we discuss several applications and how they can benefit from accounting for the physical nature of scenes (see Figure 1 for examples).

3D Reconstruction: Growing popularity of 3D displays is leading to growing demand to create 3D models of scenes from photographs. From a single image, 3D reconstruction requires interpreting the pixels of the image as surfaces with depths and orientations. Often, perfect depth estimates are not necessary; a rough sense of geometry combined with texture mapping provides convincing detail. In Chapter 5, we will discuss two approaches for 3D reconstruction from a single image.

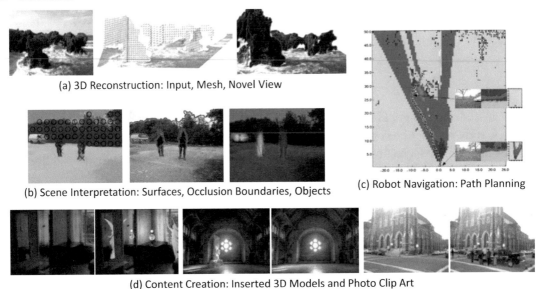

(a) 3D Reconstruction: Input, Mesh, Novel View

(b) Scene Interpretation: Surfaces, Occlusion Boundaries, Objects

(c) Robot Navigation: Path Planning

(d) Content Creation: Inserted 3D Models and Photo Clip Art

Figure 1: Several applications that require 3D scene interpretation from images. (a) Automatic 3D reconstruction from a single image; (b) Interpretation of 3D surface orientations, occlusion boundaries, and detected objects; (c) Use of surface layout for long-range path planning of a robot; (d) Insertion of objects from 3D models (left two scenes) or image regions (right two scenes), based on inferred perspective and lighting. Figures excerpted from: (a) [99]; (b) [103]; (c) [164]; (d, left) Kevin Karsch, unpublished; (d, right) [130].

Robot Navigation: We are getting closer and closer to being able to create household robots and cars that drive themselves. Such applications require finding paths, detecting obstacles, and predicting other objects' movements within the scene space. Mobile robots and autonomous vehicles will likely have additional depth sensors, but image analysis is required for long-range planning and for integrating information across multiple sensors. Most important, we need to develop good abstractions for interaction and prediction in the physical space.

Scene Interpretation: For applications such as security, robotics, and multimedia organization, we need to interpret the scene from images. We want, not just to list the objects, but to infer their goals and interactions with other objects and surfaces. This requires the ability to estimate spatial layout, occlusion boundaries, object pose, shadows, light sources, and other scene elements.

Content Creation: Every day, people around the world capture millions of snapshots of life on digital cameras and cell phones. But sometimes we want to create our own reality. We want to remove that ex-boyfriend from the Paris photo and fill in the background or to add a goofy alien into our living room for the Christmas photo or to depict Harry Potter shooting a glowing Petronus at a specter. In all of these cases, our job is much easier if we can interpret the images in terms of occlusion boundaries, spatial layout, physical lighting, and other scene properties. For example, to add a 3D model of an alien into the photograph, we need to know what size and perspective to render, how the object should be lit, and how the shading will change on nearby surfaces.

WHY WE SHOULD MODEL OBJECTS IN 3D

We want to recognize objects so that we know how to interact with them, can predict their movements, and can describe them to others. For example, we want to know which way an object is pointing and how to pick it up. Such knowledge requires some sense of the 3D layout of the objects. Spatial layout could be represented with a large number of 2D views, a full 3D model, or something in-between. Object representations that more explicitly account for the 3D nature of objects have the potential for interpolation between viewpoints, more compact representations, and better generalization for new viewpoints and poses. 3D models of objects are also crucial for 3D reconstruction, object grasping, and multi-view recognition, and 3D models are helpful for tracking and reasoning about occlusion.

OVERVIEW OF BOOK

As outlined below, this book aims to provide an understanding of important concepts in 3D scene and object representation and inference from still images. We focus on recent efforts to fuse models of geometry and perspective with statistical learning. We do not cover some of the more traditional approaches to geometric scene modeling, such as structure-from-motion, stereoscopic images, laser range finders, and structured light, which tend to draw on different techniques. Yet, even within our limited scope, we cannot describe all related work in detail. Instead, we introduce each of the major topics with a broad representative overview of related work, followed by discussion of key concepts and techniques, and concluded with a more detailed description of a few sample approaches. The detailed samples, often drawn from our own work, are illustrative but not fully representative of the important efforts by many research groups. We encourage the reader to explore beyond this book and hope that some will be inspired to forge new frontiers in 3D scene interpretation and object recognition.

Part I: Interpretation of Physical Space from an Image
- *Chapter 2: Background on 3D Scene Models*: Provides a brief overview of theories and representations for 3D scene understanding and representation.
- *Chapter 3: Single-view Geometry*: Introduces basic concepts and mathematics of perspective and single-view geometry.
- *Chapter 4: Modeling the Physical Scene*: Catalogues the elements of scene understanding and organizes the representations of physical scene space.
- *Chapter 5: Categorizing Images and Regions*: Summarizes concepts in machine learning and feature representations needed to learn models of appearance from images.

- *Chapter 6: Examples of 3D Scene Interpretation*: Describes three approaches to interpret 3D scenes from an image, illustrating some of the ideas from earlier chapters.

Part II: Recognition of 3D Objects from an Image

- *Chapter 7: Background on 3D Recognition*: Provides a brief overview of theories and representations for 3D objects.
- *Chapter 8: Modeling 3D Objects*: Reviews recently proposed computational models for 3D object recognition and categorization.
- *Chapter 9: Recognizing and Understanding 3D Objects*: Discusses datasets and machine learning techniques for 3D object recognition.
- *Chapter 10: Examples of 2D 1/2 Layout Models*: Describes two approaches to represent objects, learn the appearance models, and recognize them in novel images, exemplifying the ideas from the earlier chapters.

Part III: Integrated 3D Scene Interpretation

- *Chapter 11: Reasoning about Objects and Scenes*: Discusses and provides examples of the relations of objects and scenes can be modeled and used to improve accuracy and consistency of interpretation.
- *Chapter 12: Cascades of Classifiers*: Describes a simple sequential approach to synergistically combining multiple scene understanding tasks, with two detailed examples.
- *Chapter 13: Conclusion and Future Directions*: Briefly discusses a few important directions for future work.

Derek Hoiem and Silvio Savarese
July 2011

Acknowledgments

Derek Hoiem was supported, in part, by NSF Awards 09-04209 and 09-16014 during the writing of this book, and by a gift from Microsoft. Derek thanks his collaborators in some of the discussed original works, particularly Alexei A. Efros, Martial Hebert, Varsha Hedau, David A. Forsyth, and Kevin Karsch.

Silvio Savarese was supported by an NSF CAREER, NSF EAGER and the Gigascale Systems Research Center during the writing of this book. Silvio wishes to thank his collaborators and in particular Fei-Fei Li, David A. Forsyth, and Benjamin Kuipers for helpful feedback.

Silvio and Derek thank the many researchers who offered suggestions for the text or contributed figures.

Derek Hoiem and Silvio Savarese
July 2011

Figure Credits

Figure 1.1a, b
from [98], Hoiem, *Seeing the World Behind the Image: Spatial Layout for 3D Scene Understanding*, 2007, and from Hoiem et al, Closing the loop on scene interpretation, 2008. Copyright © 2008 IEEE. Used with permission.

Figure 1.1c
from [164], Nabbe et al, Opportunistic use of vision to push back the path-planning horizon. Copyright © 2006 IEEE. Used with permission. DOI: 10.1109/IROS.2006.281676

Figure 1.1d (r)
from [130], Lalonde et al, Photo clip art. Copyright © 2007, Association for Computing Machinery. Reprinted by permission.

Figure 2.1
from [16], Biederman et al, On the semantics of a glance at a scene, in Kubovy and Pomerantz, eds., *Perceptual Organization*, Chapter 8. Copyright © 1981 Lawrence Erlbaum. Used with permission of Taylor and Francis Group.

Figure 2.2
from [174], Ohta et al: An Analysis system for scenes containing objects with substructures. Copyright © 1978 IEEE. Used with permission.

Figure 2.5
illustration courtesy of Steve Seitz.

Figure 3.1
from [37], Criminisi, et al: *Accurate Visual Metrology from Single and Multiple Uncalibrated Images,* Copyright © 2001 Springer-Verlag. Used with the kind permission of Springer Science+Business Media. DOI: 10.1023/A:1026598000963

Figure 4.1a
from [176], Oliva and Torralba, Building the gist of a scene: the role of global image features in recognition. Copyright © 2006 Elsevier. Used with permission. DOI: 10.1016/S0079-6123(06)55002-2

Figure 4.1b
from [166], Nedovic et al: Depth information by stage classification. Copyright ©2007 IEEE. Used with permission. DOI: 10.1109/ICCV.2007.4409056

Figure 4.1c
from [102], Hoiem et al, Recovering surface layout from an image, 2007. Copyright © 2007 Springer. Used with the Kind permission of Springer Science+Business Media. DOI: 10.1007/s11263-006-0031-y

Figure 4.1d
from [196], Weiss, Yair, Bernhard Schölkopf, and John Platt, eds., *Advances in Neural Information Processing Systems 18: Proceedings of the 2005 Conference*, figure from Saxena, Chung, and Ng article, © 2006 Massachusetts Institute of Technology, by permission of The MIT Press.

Figure 4.1e
from [2], Agarwal et al, Building Rome in a Day, 2009. Copyright © 2009 IEEE. Used with permission.

Figure 4.1h
from [133], Lee et al, Geometric reasoning for single image structure recovery. Copyright © 2009 IEEE. Used with permission. DOI: 10.1109/CVPRW.2009.5206872

Figure 4.1j
from [84], Gupta et al, Blocks world revisited: Image understanding using qualitative geometry and mechanics, 2010. Copyright © 2010 Springer. Used with the Kind permission of Springer Science+Business Media. DOI: 10.1007/978-3-642-15561-1

Figure 4.1k
from [93], Hedau et al, Recovering the spatial layout of cluttered rooms, 2009. Copyright © 2009 IEEE. Used with permission. DOI: 10.1109/ICCV.2009.5459411

Figure 6.1, 6.1, and 6.3
from [102], Hoiem et al, Recovering surface layout from an image, 2007. Copyright © 2007 Springer. Used with the Kind permission of Springer Science+Business Media DOI: 10.1007/s11263-006-0031-y

Figures 6.4, 6.5, and 6.6
from [99], Hoiem et al: Automatic photo pop-up. In *ACM SIGGRAPH*, 2005. Copyright © 2005, Association for Computing Machinery. Reprinted by permission. DOI: 10.1145/1186822.1073232

Figure 6.8
from [197], Saxena et al, 3-d depth reconstruction from a single still image. Copyright © 2007 Springer. Used with the Kind permission of Springer Science+Business Media. DOI: 10.1007/s11263-007-0071-y

Figures 6.9, 6.10, and 6.11
from [93], Hedau et al, Recovering the spatial layout of cluttered rooms, 2009. Copyright © 2009 IEEE. Used with permission. DOI: 10.1109/ICCV.2009.5459411

Figure 7.2 panel a,
courtesy of Irving Biederman.

Figure 7.2 panel b,
courtesy of Peter Kovesi.

Figure 8.1a
from [152], Lowe: Object recognition from local scale-invariant features. Copyright © 1999 IEEE. Used with permission. DOI: 10.1109/ICCV.1999.790410

Figure 8.1b
from [190], Rothganger et al, 3d object modeling and recognition using local affine-invariant image descriptors and multi-view spatial constraints. Copyright © 2006 Springer. Used with the Kind permission of Springer Science+Business Media. DOI: 10.1007/s11263-005-3674-1

Figure 8.2a
from [232], Thomas et al, Towards multi-view object class detection. Copyright © 2006 IEEE. Used with permission. DOI: 10.1109/CVPR.2006.311

Figure 8.2b
from [29], Chiu et al, Virtual training for multi-view object class recognition, 2007. Copyright © 2007 IEEE. Used with permission. And from [30], Chiu et al, Class-specific grasping of 3d objects from a single 2d image. Copyright © 2010 IEEE. Used with permission. DOI: 10.1109/CVPR.2007.383044 And from [231], Thomas et al, Using recognition to guide a robots attention. Copyright © RSS, 2008.

Figure 8.2c
from [127], Kushal et al: Flexible object models for category-level 3d object recognition. Copyright © 2007 IEEE. Used with permission. DOI: 10.1109/CVPR.2007.383149

And from [145], Liebelt et al, Multi-view object class detection with a 3d geometric model. Copyright © 2010 IEEE. Used with permission. DOI: 10.1109/CVPR.2010.5539836 From [146], Liebelt et al, Viewpoint-independent object class detection using 3d feature maps. Copyright © 2008 IEEE. Used with permission.
DOI: 10.1109/CVPR.2008.4587614

Figure 8.2d
from [263], Yan et al: 3d model based object class detection in an arbitrary view. Copyright © 2007 IEEE. Used with permission. DOI: 10.1109/ICCV.2007.4409042

Figure 8.3a
from [106], Hoiem et al: 3d layoutcrf for multi-view object class recognition and segmentation. Copyright © 2007 IEEE. Used with permission. DOI: 10.1109/CVPR.2007.383045

Figure 9.1a
from [194], Savarese and Fei-Fei: 3D generic object categorization, localization and pose estimation. Copyright © 2007 IEEE. Used with permission.
DOI: 10.1109/ICCV.2007.4408987

Figure 9.1b
from [178], Ozuysal et al: Pose estimation for category specific multiview object localization. Copyright © 2009 IEEE. Used with permission.
DOI: 10.1109/CVPRW.2009.5206633

Figure 10.1
from [193], Savarese and Fei-Fei: View synthesis for recognizing unseen poses of object classes. Copyright © 2008 Springer. Used with permission.
DOI: 10.1109/ICCV.2007.4408897

Figure 10.1, 10.4, 10.5, 10.6, 10.7, 10.8, 10.9
from [218], Su et al.: Learning a dense multi-view representation for detection, viewpoint classification and synthesis of object categories. Copyright © 2009 IEEE. Used with permission. DOI: 10.1109/ICCV.2009.5459168

Figures 11.1-11.4
from [98], Hoiem, *Seeing the World Behind the Image: Spatial Layout for 3D Scene Understanding*, 2007. Also published in [106], Hoiem et al: 3d layoutcrf for multi-view object class recognition and segmentation. Copyright © 2007 IEEE. Used with permission.
DOI: 10.1109/CVPR.2007.383045

Figure 11.5
from [94], Hedau et al, Thinking inside the box: Using appearance models and context based on room geometry. Copyright © 2010 Springer. Used with the Kind permission of Springer Science+Business Media.

Figures 12.1, 12.2, 12.4, 12.6, 12.7, and 12.8
from [103], Hoiem et al: Closing the loop on scene interpretation. Copyright © 2008 IEEE. Used with permission. DOI: 10.1109/CVPR.2008.4587587

Figure 12.9
from [141], Li et al: Towards holistic scene understanding: Feedback enabled cascaded classification models, in *NIPS*, 2010. Courtesy of the authors.

Figure 12.11
from [142], Li et al: Towards total scene understanding: Classification, annotation, and segmentation in an automatic framework. Copyright © 2009 IEEE. Used with permission.

PART I

Interpretation of Physical Space from an Image

3D scene understanding has long been considered the grand challenge of computer vision. In Chapter 1, we survey some of the theories and approaches that have been developed by cognitive scientists and artificial intelligence researchers over the past several decades. To even begin to figure out how to make computers interpret the physical scene from an image, we need to understand the consequences of projection and the basic math behind projective geometry, which we lay out in Chapter 2. In Chapter 3, we discuss some of the different ways that we can model the physical scene and spatial layout. To apply scene models to new images, most current approaches train classifiers to predict scene elements from image features. In Chapter 4, we provide some guidelines for choosing features and classifiers and summarize some broadly useful image cues. Finally, in Chapter 5, we describe several approaches for recovering spatial layout from an image, drawing on the ideas discussed in the earlier chapters.

C H A P T E R   1

Background on 3D Scene Models

The beginning of knowledge is the discovery of something we do not understand.
Frank Herbert (1920–1986)

The mechanism by which humans can perceive depth from light is a mystery that has inspired centuries of vision research, creating a vast wealth of ideas and techniques that make our own work possible. In this chapter, we provide a brief historical context for 3D scene understanding, ranging from early philosophical speculations to more recent work in computer vision.

1.1   THEORIES OF VISION

The elementary impressions of a visual world are those of surface and edge.
James Gibson, *Perception of a Visual World* (1950)

For centuries, scholars have pondered the mental metamorphosis from the visual field (2D retinal image) to the visual world (our perception of 3D environment). In 1838, Charles Wheatstone [256] offered the explanation of stereopsis, that "the mind perceives an object of three dimensions by means of the two dissimilar pictures projected by it on the two retina." Wheatstone convincingly demonstrated the idea with the stereoscope but noted that the idea applies only to nearby objects.

Hermann von Helmholtz, the most notable of the 19th century empiricists, believed in an "unconscious inference", that our perception of the scene is based, not only on the immediate sensory evidence, but on our long history of visual experiences and interactions with the world [252]. This inference is based on an accumulation of evidence from a variety of cues, such as the horizon, shadows, atmospheric effects, and familiar objects.

Gibson laid out [76] a theory of visual space perception which includes the following tenets: 1) the fundamental condition for seeing the visible world is an array of physical surfaces, 2) these surfaces are of two extreme types: frontal and longitudinal, and 3) perception of depth and distance is reducible to the problem of the perception of longitudinal surfaces. Gibson further theorized that gradients are the mechanism by which we perceive surfaces.

By the 1970s, several researchers had become interested in computational models for human vision. Barrow and Tenenbaum proposed the notion of *intrinsic images* [12], capturing characteristics, such as reflectance, illumination, surface orientation, and distance, that humans are able to recover from an image under a wide range of viewing conditions. Meanwhile, David Marr proposed a three-stage theory of human visual processing [157]: from primal sketch (encoding edges and regions boundaries), to $2\frac{1}{2}$D sketch (encoding local surface orientations and discontinuities), to the full 3D model representation.

1.1.1   DEPTH AND SURFACE PERCEPTION

The extent and manner in which humans are able to estimate depth and surface orientations has long been studied. Although the debate continues on nearly every aspect of recovery and use of spatial layout, some understanding is emerging. Three ideas in particular are highly relevant: 1) that monocular cues provide much of our depth and layout sensing ability; 2) that an important part of our layout representation and reasoning primarily is based on surfaces, rather than metric depth; and 3) that many scene understanding processes depend on the viewpoint.

Cutting and Vishton [38], using a compilation of experimental data where available and logical analysis elsewhere, rank visual cues according to their effectiveness in determining ordinal depth relationships (which object is closer). They study a variety of monocular, binocular, and motion cues. They define three spatial ranges: personal space (within one or two meters), action space (within thirty meters), and vista space (beyond thirty meters) and find that the effectiveness of a particular cue depends largely on the spatial range. Within the personal space, for instance, the five most important cues primarily correspond to motion and stereopsis: interposition,[1] stereopsis, parallax, relative size, and accommodation. In the vista range, however, all of the top five cues are monocular: interposition, relative size, height in visual field, relative density, and atmospheric perspective. Likewise, in the action range, three of the top five most important cues are monocular. Cutting and Vishton suggest that stereo and motion cues have been overemphasized due to their importance in the personal space and that much more study is needed in the monocular cues that are so critical for spatial understanding of the broader environment.

Koenderink and colleagues [120, 121] experimentally measure the human's ability to infer depth and local orientation from an image. Their subjects were not able to accurately (or even consistently) estimate depths of a 3D form (e.g., a statue), but could indicate local surface orientations. They further provide evidence that people cannot determine the relative depth of two points unless there is some visible and monotonic surface that connects them. Zimmerman et al. [272] provide additional evidence that people can make accurate slant judgements but are unable to accurately estimate depth of disconnected surfaces. These experimental results confirm the intuitions of Gibson and others – that humans perceive the 3D scene, not in terms of absolute depth maps, but in terms of surfaces.

---

[1]Note that interposition provides near-perfect ordinal depth information, since the occluding object is always in front, but is uninformative about metric depth.

Figure 1.1: Examples violations of Biederman's five relational constraints. On the left is a violation of position; in the center, a violation of interposition; and on the right, there are violations of support, size, and probability of appearance. Illustration from [16], used with permission.

There has been an enormous amount of research into the extent to which scene representations and analysis are dependent on viewpoint. If human visual processing is viewpoint-independent, the implication is that we store 3D models of scenes and objects in our memory and manipulate them to match the current scene. Viewpoint-independence, however, implies that the retinotopic image must somehow be registered to the 3D model. In a viewpoint-dependent representation, different models are stored for different views. While it is likely that both viewpoint-dependent and viewpoint-independent storage and processing exist, the evidence indicates that a view-dependent representation is more dominant. For instance, Chua and Chun [31] find that implicit scene encodings are viewpoint dependent. Subjects were able to find targets faster when presented with previously viewed synthetic 3D scenes, but the gain in speed deteriorates as the viewpoint in a repeated viewing differs from the original. Gottesman [78] provides further evidence with experiments involving both photographs and computer-generated images. The subjects are first primed with an image of the scene and then presented with the target, which is the same scene rotated between 0 and 90 degrees. When asked to judge the distance between two objects in the target scene, response times increased with larger differences in viewpoint between the prime and the target. See Chapter 6 for further discussion of theories of 3D object recognition by humans.

1.1.2   A WELL-ORGANIZED SCENE

Irving Biederman [16] characterizes a well-organized scene as one that satisfies five relational constraints: support, interposition, probability of appearance, position, and size. See Figure 1.1 for illustrations. In order, these properties are described as follows: objects tend to be supported by a solid surface; the occluding object tends to obscure the occluded object; some objects are more likely to appear in certain scenes than others; objects often belong in a particular place in the scene; and objects usually have a small range of possible sizes. Although he does not explain how these relationships are perceived, Biederman supplies evidence that they are extremely valuable for scene understanding. For instance, when a subject is asked to determine whether a particular object is

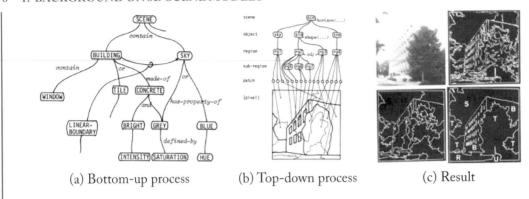

(a) Bottom-up process   (b) Top-down process   (c) Result

Figure 1.2: A system developed in 1978 by Ohta et al. [173, 174] for knowledge-based interpretation of outdoor natural scenes. The system is able to label an image (c) into semantic classes: S-sky, T-tree, R-road, B-building, U-unknown. Figure used with permission.

present in a scene, the response time increases when the object violates one of the relational constraints listed above. Hollingworth and Henderson [108] questioned these results (particularly for probability of appearance), showing that experimental biases could account for some of the reported differences. Instead, they suggest that people may be more likely to remember unusual or unexpected objects, possibly due to memory or attentional processes responding to surprise. This consistency effect, that humans tend to remember objects that are somehow inconsistent with their environment, has long been studied (e.g., [97, 182]) and is further evidence that such relational constraints play an important role in recognition and scene understanding.

1.2   EARLY COMPUTER VISION AND AI

Given a single picture which is a projection of a three-dimensional scene onto the two-dimensional picture plane, we usually have definite ideas about the 3-D shapes of objects. To do this we need to use assumptions about the world and the image formation process, since there exist a large number of shapes which can produce the same picture.
Takeo Kanade, "Recovery of the Three-Dimensional Shape of an Object from a Single View" (1981)

In its early days, computer vision had but a single grand goal: to provide a complete semantic interpretation of an input image by reasoning about the 3D scene that generated it. Some of this effort focused on achieving 3D reconstructions of objects or simple scenes, usually from line drawings. For example, Kanade [117] demonstrates how assumptions about geometric properties such as planarity, parallelism and skewed symmetry can be used to reconstruct the 3D shape of a chair. Barrow and Tenenbaum [13] suggest that contour information should be combined with other cues, such as texture gradient, stereopsis, and shading and speculate on the primacy of geometric cues in early vision.

Initial attempts at a broader scene understanding focused largely on toy "blocks worlds" [86, 188], but, by the 1970s, several extremely sophisticated approaches were proposed for handling real indoor and outdoor images. For instance, Yakimovsky and Feldman [262] developed a Bayesian framework for analyzing road scenes that combined segmentation with semantic domain information at the region and inter-region level. Tenenbaum and Barrow proposed *Interpretation-Guided Segmentation* [230] which labeled image regions, using constraint propagation to arrive at a globally consistent scene interpretation. Ohta et al. [173, 174] combined bottom-up processing with top-down control for semantic segmentation of general outdoor images. Starting with an oversegmentation, the system generated "plan images" by merging low-level segments. Domain knowledge was represented as a semantic network in the bottom-up process (Figure 1.2a) and as a set of production rules in the top-down process (Figure 1.2b). Results of applying this semantic interpretation to an outdoor image are shown on Figure 1.2c. By the late 1970s, several complete image understanding systems were being developed including such pioneering work as Brooks' *ACRONYM* [24] and Hanson and Riseman's *VISIONS* [89]. For example, *VISIONS* was an ambitious system that analyzed a scene on many interrelated levels including segments, 3D surfaces and volumes, objects, and scene categories.

It is interesting to note that a lot of what are considered modern ideas in computer vision—region and boundary descriptors, superpixels, combining bottom-up and top-down processing, Bayesian formulation, feature selection, etc.—were well known three decades ago. But, although much was learned in the development of these early systems, it was eventually seen that the hand-tuned algorithms could not generalize well to new scenes.

This, in turn, leads people to doubt the very goal of complete image understanding. However, it seems that the early pioneers were simply ahead of their time. They had no choice but to rely on heuristics because they lacked the large amounts of data and the computational resources to *learn* the relationships governing the structure of our visual world.

1.3   MODERN COMPUTER VISION

The failures of early researchers to provide robust solutions to many real-world tasks led to a new paradigm in computer vision: rather than treat the image as a projection from 3D, why not simply analyze it as a 2D pattern? Statistical methods and pattern recognition became increasingly popular, leading to breakthroughs in face recognition, object detection, image processing, and other areas. Success came from leveraging modern machine learning tools, large data sets, and increasingly powerful computers to develop data-driven, statistical algorithms for image analysis.

It has become increasingly clear, however, that a purely 2D approach to vision cannot adequately address the larger problem of scene understanding because it fails to exploit the relationships that exist in the 3D scene. Several researchers have responded, and much progress in recent years has been made in spatial perception and representation and more global methods for reasoning about the scene.

Song-Chun Zhu and colleagues have contributed much research in computational algorithms for spatial perception and model-driven segmentation. Guo et al. [81] propose an implementation of Marr's primal sketch, and Han and Zhu [88] describe a grammar for parsing image primitives. Tu and Zhu [244] describe a segmentation method based on a generative image representation that could be used to sample multiple segmentations of an image. Barbu and Zhu [9] propose a model-driven segmentation with potential applications to image parsing [243].

Nitzberg and Mumford [170] describe a 2.1D sketch representation, a segmentation and layering of the scene, and propose an algorithm for recovering the 2.1-D sketch from an image. The algorithms consists of three phases: finding edges and T-junctions, hypothesizing edge continuations, and combinatorially finding the interpretation that minimizes a specified energy function. They show that the algorithm can account for human interpretations of some optical illusions. Huang, Lee, and Mumford [112] investigate the statistics of range images, which could help provide a prior on the 3D shape of the scene.

Oliva and Torralba [175, 176] characterize the "spatial envelope" of scenes with a set of continuous attributes: naturalness, openness, roughness, expansion, and ruggedness. They further provide an algorithm for estimating these properties from the spectral signature of an image and, in [239], used similar algorithms to estimate mean depth of the scene. These concepts have since been applied to object recognition [162, 219] and recovery of depth information based on objects in an image [220].

Other recent works have attempted to model the 3D scene using a wide variety of models, including: a predefined set of prototype global scene geometries [166]; a *gist* [175] of a scene describing its spatial characteristics; a 3D box [93, 109, 207] or collection of 3D polyhedrals [87, 177]; boundaries between ground and walls [10, 42, 133]; depth-ordered planes [268]; a pixel labeling of approximate local surface orientations [102], possibly with ordering constraints [107, 148]; a blocks world [84]; or depth estimates at each pixel [197].

In the following three chapters, we discuss many ways to model the space of scenes and relevant techniques and features to estimate them from images, including a detailed discussion of three representative approaches.

CHAPTER 2

Single-view Geometry

When we open an eye or take a photograph, we see only a flattened, two-dimensional projection of the physical underlying scene. The consequences are numerous and startling. Size relationships are distorted, right angles are suddenly wrong, and parallel lines now intersect. Objects that were once apart are now overlapping in the image, so that some parts are not visible. At first glance, it may seem that the imaging process has corrupted a perfectly reasonable, well-ordered scene into a chaotic jumble of colors and textures. However, with careful scene representation that accounts for perspective projection, we can tease out the structure of the physical world.

In this chapter, we summarize the consequences of the imaging process and how to mathematically model perspective projection. We will also show how to recover an estimate of the ground plane, which allows us to recover some of the relations among objects in the scene.

2.1 CONSEQUENCES OF PROJECTION

The photograph is a projection of the 3D scene onto a 2D image plane. Most cameras use a **perspective projection**. Instead of recording the 3D position of objects (X, Y, Z), we observe their projection onto the 2D image plane at (u, v). In this projection, a set of parallel lines in 3D will intersect at a single point, called the **vanishing point**. A set of parallel planes in 3D intersect in a single line, called the **vanishing line**. The vanishing line of the ground plane, called the **horizon**, is of particular interest because many objects are oriented according to the ground and gravity. See Figure 2.1 for an illustration of vanishing points and lines.

Another major consequence of projection is **occlusion**. Because light does not pass through most objects, only the nearest object along a ray is visible. Take a glance at the world around you. Every object is occluding something, and most are partially occluded themselves. In some ways, occlusion simplifies the scene interpretation. Imagine if you could see every object for miles in a single projection! In other ways, occlusion makes scene interpretation more difficult by hiding parts and disturbing silhouettes.

2.2 PERSPECTIVE PROJECTION WITH PINHOLE CAMERA: 3D TO 2D

We often assume a **pinhole camera** to model projection from 3D to 2D, as shown in Figure 2.2. Under this model, light passes from the object through a small pinhole onto a sensor. The pinhole model ignores lens distortion and other nonlinear effects, modeling the camera in terms of its intrinsic

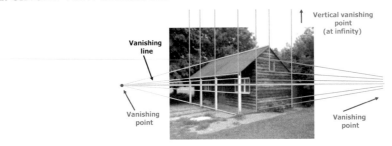

Figure 2.1: Illustration of vanishing points and lines. Lines that are parallel in 3D intersect at a point in the image, called the *vanishing point*. Each set of parallel lines that is parallel to a plane will intersect on the *vanishing line* of that plane. The vanishing line of the ground plane is called the *horizon*. If a set of parallel lines is also parallel to the image plane, it will intersect at infinity. Three finite orthogonal vanishing points can be used to estimate the intrinsic camera parameters. Figure from [37].

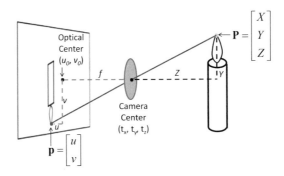

Figure 2.2: Illustration of pinhole camera model. The camera, with focal length f and optical center (u_0, v_0), projects from a 3D point in the world to a 2D point on the image plane.

parameters (focal length f, optical center (u_0, v_0), pixel aspect ratio α, and skew s) and extrinsic parameters (3D rotation $\mathbf{R}$ and translation $\mathbf{t}$). Although the image plane is physically behind the pinhole, as shown in the figure, it is sometimes convenient to pretend that the image plane is in front, with focal length f, so that the image is not inverted. If we assume no rotation ($\mathbf{R}$ is identity), and camera translation of $(0, 0, 0)$, with no skew and unit aspect ratio, we can use the properties of similar triangles to show that $u - u_0 = f\frac{X}{Z}$ and $v - v_0 = f\frac{Y}{Z}$.

We can more easily write the projection as a system of linear equations in **homogeneous coordinates**. These homogeneous coordinates add a scale coordinate to the Cartesian coordinates, making them convenient for representing rays (as in projection) and direction to infinitely distant points (e.g., where 3D parallel lines intersect). To convert from Cartesian to homogeneous coordinates, append a value of 1 (e.g., $(u, v) \rightarrow (u, v, 1)$ and $(X, Y, Z) \rightarrow (X, Y, Z, 1)$). To convert from homogeneous to Cartesian coordinates, divide by the last coordinate (e.g., $(u, v, w) \rightarrow (u/w, v/w)$).

Under homogeneous coordinates, our model of pinhole projection from 3D world coordinates (X, Y, Z) to image coordinates (u, v) is written as follows:

$$
\begin{bmatrix} w \cdot u \\ w \cdot v \\ w \end{bmatrix} = \mathbf{K} \begin{bmatrix} \mathbf{R} & \mathbf{t} \end{bmatrix} \begin{bmatrix} X \\ Y \\ Z \\ 1 \end{bmatrix} \quad \text{or} \quad \begin{bmatrix} w \cdot u \\ w \cdot v \\ w \end{bmatrix} = \begin{bmatrix} f & s & u_0 \\ 0 & \alpha f & v_0 \\ 0 & 0 & 1 \end{bmatrix} \begin{bmatrix} r_{11} & r_{12} & r_{13} & t_x \\ r_{21} & r_{22} & r_{23} & t_y \\ r_{31} & r_{32} & r_{33} & t_z \end{bmatrix} \begin{bmatrix} X \\ Y \\ Z \\ 1 \end{bmatrix}.
$$
$$(2.1)$$

Note that the intrinsic parameter matrix $\mathbf{K}$ has only five parameters and that the rotation and translation each have three parameters, so that there are 11 parameters in total. Even though the rotation matrix has nine elements, each element can be computed from the three angles of rotation. We commonly assume unit aspect ratio and zero skew as a good approximation for most modern cameras. Sometimes, it is also convenient to define world coordinates according to the camera position and orientation, yielding the simplified

$$
\begin{bmatrix} w \cdot u \\ w \cdot v \\ w \end{bmatrix} = \begin{bmatrix} f & 0 & u_0 \\ 0 & f & v_0 \\ 0 & 0 & 1 \end{bmatrix} \begin{bmatrix} X \\ Y \\ Z \end{bmatrix}.
$$
$$(2.2)$$

>From this equation, we get the same result as we found using similar triangles in Figure 2.2. In Cartesian coordinates, $u = f\frac{X}{Z} + u_0$ and $v = f\frac{Y}{Z} + v_0$.

2.3   3D MEASUREMENT FROM A 2D IMAGE

As illustrated in Figure 2.3, an infinite number of 3D geometrical configurations could produce the same 2D image, if only photographed from the correct perspective. Mathematically, therefore, we have no way to recover 3D points or measurements from a single image. Fortunately, because our world is so structured, we *can* often make good estimates. For example, if we know how the camera is rotated with respect to the ground plane and vertical direction, we can recover the relative 3D heights of objects from their 2D positions and heights.

Depending on the application, the world coordinates can be encoded using either three orthogonal vanishing points or the projection matrix, leading to different approaches to 3D measurement. When performing estimation from a single image, it is helpful to think in terms of the horizon line (the vanishing line of the ground plane) and the vertical vanishing point.

For example, look at the 3D scene and corresponding image in Figure 2.4. Suppose the camera is level with the ground, so that the vertical vanishing point is at infinity and the horizon is a line through row v_h. Then, if a grounded object is as tall as the camera is high, the top of the object will be projected to v_h as well. We can use this and basic trigonometry to show that $\frac{Y_o}{Y_c} = \frac{v_t - v_b}{v_h - v_b}$, where Y_o is the 3D height at the top of the object (defining the ground at $Y = 0$), Y_c is the camera

Original Image   Interpretation 1: Painting on Ground

Interpretation 2: Floating Objects   Interpretation 3: Man in Field

Figure 2.3: Original image and novel views under three different 3D interpretations. Each of these projects back into the input image from the original viewpoint. Although any image could be produced by any of an infinite number of 3D geometrical configurations, very few are plausible. Our main challenge is to learn the structure of the world to determine the most likely solution.

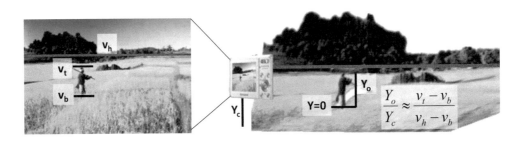

$$\frac{Y_o}{Y_c} \approx \frac{v_t - v_b}{v_h - v_b}$$

Figure 2.4: The measure of man. Even from one image, we can measure the object's height, relative to the camera height, if we can see where it contacts the ground.

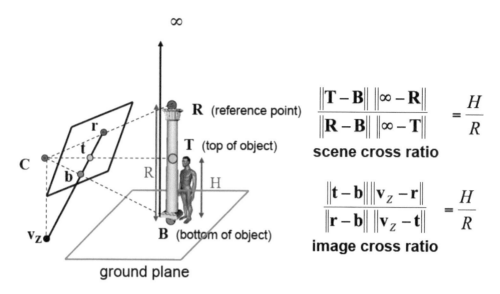

Figure 2.5: This figure by Steve Seitz illustrates the use of the cross-ratio to make world measurements from an image.

height, and v_t and v_b are the top and bottom positions of the object in the image. This relationship is explored in more detail in Hoiem et al. [104] and used to aid in object recognition, a topic we will discuss in a later chapter.

As shown by Criminisi et al. [37], the **cross ratio invariant** can be used to relate the 3D heights of objects under more general conditions. The cross ratio of 4 collinear points (e.g., $\frac{\|\mathbf{p}_3 - \mathbf{p}_1\| \|\mathbf{p}_4 - \mathbf{p}_2\|}{\|\mathbf{p}_3 - \mathbf{p}_2\| \|\mathbf{p}_4 - \mathbf{p}_1\|}$, for any collinear points $\mathbf{p}_{1..4}$) does not change under projective transformations. If one of those points is a vanishing point, then its 3D position is at infinity, simplifying the ratio. Using the vertical vanishing point, we can recover the relative heights of objects in the scene (see Figure 2.5 for an illustration).

We have a special case when the camera is upright and not slanted so that the vertical vanishing point is at infinity in the image plane. In this case, $v_Z = \infty$, so we have $\frac{Y_t - Y_b}{Y_c - Y_b} = \frac{v_t - v_b}{v_h - v_b}$, where Y_t is the height of the top of the object, Y_b is the height at the bottom, Y_c is the height of the camera, and v_t, v_b, and v_h, are the image row coordinates of the object and horizon.

2.4   AUTOMATIC ESTIMATION OF VANISHING POINTS

Recall that all lines with the same 3D orientation will converge to a single point in an image, called a vanishing point. Likewise, sets of planes converge to a vanishing line in the image plane. If a set of parallel lines is also parallel to a particular plane, then their vanishing point will lie on the vanishing line of the plane.

If we can recover vanishing points, we can use them to make a good guess at the 3D orientations of lines in the scene. As we will see, if we can recover a triplet of vanishing points that correspond to orthogonal sets of lines, we can solve for the focal length and optical center of the camera. Typically, two of the orthogonal vanishing points will be on the horizon and the third will be in the vertical direction. If desired, we can solve for the rotation matrix that sets the orthogonal directions as the X, Y, and Z directions.

Most vanishing point detection algorithms work by detecting lines and clustering them into groups that are nearly intersecting. If we suspect that the three dominant vanishing directions will be mutually orthogonal, e.g., as in architectural scenes, we can constrain the triplet of vanishing points to provide reasonable camera parameters. However, there are a few challenges that make the problem of recovering vanishing points robustly much more tricky than it sounds. In the scene, many lines, such as those on a tree's branches, point in unusual 3D orientations. Nearly all pairs of lines will intersect in the image plane, even if they are not parallel in 3D, leading to many spurious candidates for vanishing points. Further, inevitable small errors in line localization cause sets of lines that really are parallel in 3D not to intersect at a single point. Finally, an image may contain many lines, and it is difficult to quickly find the most promising set of vanishing points.

BASIC ALGORITHM TO ESTIMATE THREE MUTUALLY ORTHOGONAL VANISHING POINTS:

1. Detect straight line segments in an image, represented by center (u_i, v_i) and angle θ_i in radians. Ignore lines with fewer than L pixels (e.g., $L{=}20$), because these are less likely to be correct detections or to have large angular error. Kosecka and Zhang [123] describe an effective line detection algorithm (see Derek Hoiem's software page for an implementation, currently at `http://www.cs.illinois.edu/homes/dhoiem/software`).

2. Find intersection points of all pairs of lines to create a set of vanishing point candidates. It's easiest to represent points in homogeneous coordinates, $\mathbf{p}_i = [w_i u_i \quad w_i v_i \quad w_i]^T$, and lines in the form $\mathbf{l}_i = [a_i \quad b_i \quad c_i]^T$ using the line equation $ax + by + c = 0$. The intersection of lines $\mathbf{l}_i$ and $\mathbf{l}_j$ is given by their cross product: $\mathbf{p}_{ij} = \mathbf{l}_i \times \mathbf{l}_j$. If the lines are parallel, then $w_{ij} = 0$. Likewise, the line formed by two points $\mathbf{p}_i$ and $\mathbf{p}_j$ is given by their cross product: $\mathbf{l}_{ij} = \mathbf{p}_i \times \mathbf{p}_j$.

3. Compute a score s_j for each vanishing point candidate (u_j, v_j): $s_j = \sum_i |l_i| \exp\left(-\frac{|\alpha_i - \theta_i|}{2\sigma^2}\right)$, where $|l_i|$ is the length and θ_i is the orientation (in radians) of line segment i, α_i is the angle from the center of the line segment (u_i, v_i) to (u_j, v_j) in radians, and σ is a scale parameters (e.g., $\sigma = 0.1$). In the distance between angles, note that $-2\pi = 0 = 2\pi$.

4. Choose the triplet with the highest total score that also leads to reasonable camera parameters. For a given triplet, $\mathbf{p}_i, \mathbf{p}_j, \mathbf{p}_k$, the total score is $s_{ijk} = s_i + s_j + s_k$. Orthogonality constraints specify: $\left(\mathbf{K}^{-1}\mathbf{p}_i\right)^T \left(\mathbf{K}^{-1}\mathbf{p}_j\right) = 0$; $\left(\mathbf{K}^{-1}\mathbf{p}_i\right)^T \left(\mathbf{K}^{-1}\mathbf{p}_k\right) = 0$; and $\left(\mathbf{K}^{-1}\mathbf{p}_j\right)^T \left(\mathbf{K}^{-1}\mathbf{p}_k\right) =$

0, where $\mathbf{K}$ is the 3x3 matrix in Eq. 2.2 and the points are in homogeneous coordinates. With three equations and three unknowns, we can solve for u_0, v_0, and f. We consider (u_0, v_0) within the image bounds and $f > 0$ to be reasonable parameters.

The algorithm described above is a simplified version of the methods in Rother [189] and Hedau et al. [93]. Many other algorithms are possible, such as [11, 123, 226], that improve efficiency, handle vanishing points that are not mutually orthogonal, incorporate new priors, or have different ways of scoring the candidates. Implementations from Tardif [226] and Barinova et al. [11] are available online. Sometimes, for complicated or non-architectural scenes, example-based methods [104, 239] may provide better estimates of perspective than vanishing point based methods.

2.5   SUMMARY OF KEY CONCEPTS

The camera projects 3D points onto a 2D imaging plane. In consequence, parallel 3D lines converge to a vanishing point on the image plane, size relationships and angles between lines are distorted, and nearer objects occlude further objects along a ray. The true sizes, positions, and orientations of objects and surfaces can often be recovered by modeling the camera, typically with a pinhole camera model. The reader should understand and memorize the pinhole projection equation (Eq. 2.1) and the notation of the intrinsic and extrinsic camera matrices. To recover 3D size and position in an image, we need to know the distance and orientation between the camera and the supporting 3D plane, typically through vanishing point estimation or detection of known objects.

C H A P T E R   3

Modeling the Physical Scene

All models are wrong, but some are useful.
George E.P. Box and Norman R. Draper, *Empirical Model-Building and Response Surfaces*
(1987)

When we humans see a photograph, we see not just a 2D pattern of color and texture, but the world behind the image. How do we do it? Many theories have been proposed, from Gestalt emergence to Helmholtzian data-driven unconscious inference to Gibson's ecological approach. We can think of the world as a set of schema or frames, with each scene organized according to some overarching structure or as a set of local observations that somehow fits together. The brain is elegant for its robustness, not for its simplicity. More like a spaceship, with lots of redundant systems, than a race car in which non-essentials are pared away. In short, there are many ways to represent a scene and many ways to infer those representations from images. We researchers do not need to come up with the single best representation. Instead, we should propose *multiple* modes of representation and try to understand how to organize them into a functioning system that enables interpretation of physical scenes from images.

In this chapter, we consider how to model the physical scene. In Section 3.1, we separately describe many components of the scene. Then, in Section 3.2, we survey some proposed approaches to model the overall scene space.

3.1   ELEMENTS OF PHYSICAL SCENE UNDERSTANDING

Although not a perfect dichotomy, it is sometimes helpful to think of the world as composed of objects and surfaces, or, in Adelson's terms "things" and "stuff" [1]. Objects, the "things," have a finite form and tend to be countable and movable. Surfaces, the "stuff," are amorphous and uncountable like sand, water, grass, or ground. If you break off a piece of stuff, you still have stuff. If you break off a piece of an object, you have part of an object. Surfaces are the stage; objects are the players and props. To understand a 3D scene, we need to interpret the image in terms of the various surfaces and objects, estimating their visual and semantic properties within the context of the event.

3.1.1   ELEMENTS

Surfaces can be described by geometry (depth, orientation, closed shape, boundaries), material, and category. For example, we may want to know: that a particular surface is flat, horizontal, long, and rectangular; that the material absorbs little water, is very hard and rough, and is called concrete; and that the surface as a whole is called a sidewalk. Sometimes, it is helpful to decompose the pixel intensity into shading and albedo terms. The resulting shading and albedo maps are often called "intrinsic images" [225], though the original definition of intrinsic images [12] encompassed geometric properties of surfaces as well as illumination properties. Shading, a function of surface orientation and light sources, provides cues for surface orientation and shape. Albedo specifies the fraction of reflected light as a function of the wavelength (the surface brightness and color) and is a property of the material.

Objects can be described according to their surfaces (geometry and material), as well as their shapes, parts, functionality, condition, pose, and activity, among other attributes. For a particular object, we may want to find its boundaries, to identify it as an animal, and specifically a dalmatian (dog), determine that it is lying down on the sidewalk several feet away, and that it is looking towards its owner who is holding it on a leash. Object categories are useful for reference, but we also want to know what an object is doing and its significance within a particular scene.

Materials can be described by category names (mostly useful for communication) or properties, such as hardness, viscosity, roughness, malleability, durability, albedo, and reflectivity. The material properties tell us what to expect if we walk on, touch, or smell an object or surface, and how it might look under particular lighting conditions.

Boundaries are discontinuities in the depth, orientation, material, or shading of surfaces. Detecting and interpreting boundaries is an important part of scene understanding (remember that many early works were based entirely on line drawings). Discontinuities in depth, called occlusion boundaries, are particularly important because they indicate the extent of objects and their parts. An occlusion boundary can be represented with a curve and a figure/ground label, which tells which side is in front. Occlusion complicates vision by obscuring objects, but recovered occlusion boundaries provide valuable information about depth orderings. Neurological studies emphasize the fundamental role of occlusion reasoning. In macaque brains, Bakin et al. [5] find that occlusion boundaries and contextual depth information are represented in the early V2 processing area.

Junctions occur where two or more boundaries meet in an image. For example, the corner of a room is a junction of a wall-wall boundary and a wall-floor or wall-ceiling boundary. Junctions and their angles provide constraints and valuable cues for the figure/ground labels of the boundaries between surfaces.

Light sources strongly influence an object's appearance; the observed shading and color depend on the position and brightness of each light source. Face recognition researchers have developed many techniques to improve robustness to variations in illumination. For example, Basri and Jacobs show that the images of convex Lambertian surfaces under arbitrary illumination can be modeled in a nine-dimensional subspace, and Lee et al. [134], in "Nine Points of Light", showed that this subspace can be effective captured by taking nine photographs of a subject with different positions of a single light source. More generally, knowledge of light source positions and orientations can provide strong cues to surface geometry, help to reason about shadows, and account for variation in object appearance due to lighting.

Shadows can either aid or confound scene interpretation, depending on whether we model the shadows or ignore them. Shadows are created wherever an object obscures the light source, and they are an ever-present aspect of our visual experience. Psychologists have shown that humans are not sensitive to the direction and darkness of cast shadows [27], but the presence of a shadow is crucial for interpreting object/surface contact [118]. If we can detect shadows, we can better localize objects, infer object shape, and determine where objects contact the ground. Detected shadows also provide cues for lighting direction [128] and scene geometry. On the other hand, if we ignore shadows, spurious edges on the boundaries of shadows and confusion between albedo and shading can lead to mistakes in visual processing. For these reasons, shadow detection has long been considered to be a crucial component of scene interpretation (e.g., [12, 251]).

3.1.2 PHYSICAL INTERACTIONS

To understand a scene, it's not enough to break it down into objects and surfaces. We also must understand their physical interactions, such as *support, proximity, contact, occlusion,* and *containment.* We want to know when people are walking or talking together. We should be more alert if a security guard is holding a gun than if he is wearing a holstered gun. A bird in the hand is worth two in the bush.

Despite their importance, few recent efforts have been made to model the interactions of objects. As one example, Farhadi et al. [57] try to produce *<object, action, scene>* triplets that can be mapped to sentences, such as *<horse, ride, field>* mapped to "A person is riding a horse in a field." The triplets were generated by computing features from object detector responses and textural gist descriptors over the image and using them to predict individual object, action, and scene labels. Then, these predictions are matched to the predictions of labeled training data to generate triplets and sentences.

To accurately interpret physical interactions, we need to infer the physical space of the scene, to recognize objects and infer their pose, and to estimate shading. For example, if we know the geometry of the ground plane, we can determine whether a detected person is likely to be standing on the ground (more on this in Chapter 10). By detecting shadows, we can determine what surfaces are directly beneath an object. Finally, when objects interact, they often take on certain poses or appearance [59, 83, 266]. A person riding a horse looks different than a person standing next to a horse. A dog that is eating its food will take a different pose than a dog standing near its food. None of the intermediate problems of inferring scene space or object identity and pose are yet solved, but it is worthwhile to start working towards the larger problem of describing interactions.

3.2 REPRESENTATIONS OF SCENE SPACE

One of the greatest challenges in computer vision is how to organize the elements of the physical scene into a more coherent understanding. In the last several years, many scene models have been

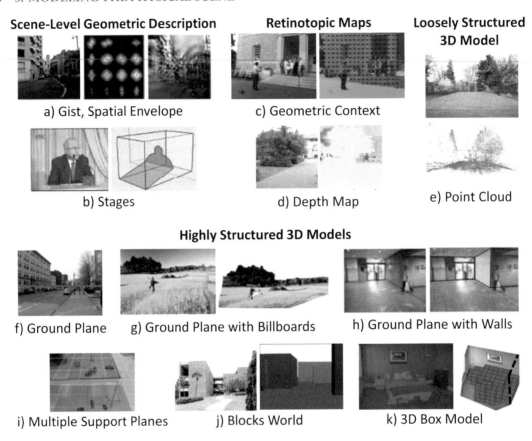

Figure 3.1: Examples of several representations of scenes. **(a)** Left: input image; Center: log-polar plots, illustrating the gist representation [175, 176] which encodes the average absolute filter responses at different orientations and scales within a 4x4 grid of spatial cells; Right: a reconstruction created by perturbing random noise to result in the same features as the input image. **(b)** Canonical 3D scenes represented as one of a set of stages [166]. **(c)** Geometric context [102], pixel labels according to several geometric classes: support (green), sky (blue), and vertical (red), with the latter subdivided into planar (left/center/right, denoted by arrows), non-planar porous ("O"), and non-planar solid ("X"). **(d)** Input image and depth map [196, 199] (yellow is close, blue is far). **(e)** Sample image and 3D point cloud computed from 457 images [2] (photo by andypick at Flickr). **(f)** Illustration of ground plane, which can be parameterized by camera height and horizon and used to put the reference frame into perspective [104]. **g)** Ground plane with vertical billboards for objects [99, 103]. **(h)** Model of an indoor scene as ground plane with vertical walls along orthogonal directions [132]. **(i)** Model in which objects can be supported by one of multiple parallel planes [8]. **(j)** Rough 3D model composed of several blocks [84]. **(k)** Left: model of the scene as a 3D box with surface labels on pixels; Right: 3d reconstruction based on estimates on left.

proposed, ranging from a rough "gist" characterization of the spatial envelope [175] to a map of the depth at each pixel [197]. See Figure 3.1 for several examples. We have a better chance of correctly estimating a very simple structure, such as a box-shaped scene, from an image, but such simple structures might not provide a good approximation for complex scenes. On the other hand, a very loose structure, such as a 3D point cloud, may be general enough to model almost any scene, but without much hope of recovering the correct parameters. Also, simpler models may provide more useful abstractions for reasoning about objects and their interactions with scene surfaces. To choose the right scene model, one must consider the application. An indoor robot could take advantage of the structure of man-made indoor scenes, while an algorithm to help a small outdoor wheeled robot to navigate might want more detailed estimates of depth and orientation.

In this section, we organize 3D scene representations into scene-level geometric descriptions, retinotopic maps, highly structured 3D models, and loosely structured 3D models. The representations range in detail from tags that characterize the scene as a whole to detailed models of geometry. The scene descriptions and retinotopic maps are highly viewpoint-dependent, while the highly structured and loosely structured 3D models are scene-centric. In the following text, we describe several examples of scene representations and discuss their usefulness.

3.2.1   SCENE-LEVEL GEOMETRIC DESCRIPTION

One approach is to define properties or categories that holistically describe the spatial layout. Oliva and Torralba [175] characterize the 3D space of the scene as "open" or "deep" or "rough". Collectively, the properties are called the "spatial envelope". This work, which was among the first data-driven attempts to glean 3D information from an image, makes a key observation that simple textural features can provide a good sense of the overall space and depth of the scene. The features, often called gist features, are measurements of the responses of frequency at multiple orientations and scales (typically measured by linear filter responses) within the cells of a spatial grid on the image. The characteristics of the spatial envelope have not been used widely in computer vision, but the underlying gist features have been used effectively for scene classification, object recognition, and geometry-based scene matching.

Nedovic et al. [166] take a different approach, assigning scenes into one of a set of geometric categories called "stages". For example, many television programs have prototypical shots, such as a person sitting behind a desk, that have consistent 3D geometry. By analyzing the image texture, the image can be assigned to one of these stages.

3.2.2   RETINOTOPIC MAPS

Going back to the intrinsic image work of Barrow and Tenenbaum [12], one of the most popular ways to represent the scene is with a set of label maps or channels that align with the image. We call these "retinotopic maps" in reference to their alignment with the retina or image sensor. In the original work, a scene is represented with a depth map, a 3D orientation map, a shading map, and an albedo map. We can create other maps for many other properties of the scene and its surfaces:

material categories, object categories, boundaries, shadows, saliency measures, and reflectivity, among others. The RGB image is a simple example, with three channels of information telling us about the magnitude of different ranges of wavelengths of light at each pixel. Such maps are convenient to create because well-understood techniques such as Markov Random Fields and region-based classification are directly applicable. The maps are also convenient to work with because they are all registered to each other. In Chapter 11, we'll see methods to provide more coherent scene interpretations by reasoning among several retinotopic maps. One significant limitation of the maps is the difficulty to encode the relations across pixels or regions or surfaces. For example, it is easy to represent which pixels correspond to the surfaces of cars, but much harder to represent that there are five cars.

3.2.3   HIGHLY STRUCTURED 3D MODELS

Because recovering 3D geometry from one 2D image is an ill-posed problem, we often need to make assumptions about the structure of the scene. If we are careful with our model designs, we can encode structures that are simple enough to infer from images and also provide useful abstractions for interpreting and interacting with the scene.

3.2.3.1   Ground Plane Model

One of the simplest and most useful scene models is that there is a ground plane that supports upright objects. The position and orientation of the ground plane with respect to the camera can be recovered from the intrinsic camera parameters (focal length, optical center), the horizon line, and the camera height. As shown in [104], if the photograph is taken from near the ground without much tilt, a good approximation can be had from just two parameters: the horizon vertical position and the camera height. Earlier (Chapter 2), we showed how to estimate an object's 3D height from its 2D coordinates using these parameters.

The ground plane model is particularly useful for object recognition in street scenes (or for nearly all scenes if you live in Champaign, IL). Street scenes tend to have a flat ground, and objects of interest are generally on the ground and upright. In Chapter 10, we will discuss an approach from Hoiem et al. [104] to detect objects and recover scene parameters simultaneously. Likewise, for object insertion in image editing, knowledge of the ground plane can be used to automatically rescale object regions as the position changes [130]. Autonomous vehicles can use the horizon position and camera parameters to improve recognition or to reduce the space that must be searched for objects and obstacles. The model may not be suitable in hilly terrain or when important objects may be off the ground, such as a bottle on a table or a cat on a couch. For non-grounded objects, simple extensions are possible to model multiple supporting planes [8].

Because the model has so few parameters, it can be accurately estimated in a wide variety of photographs. The horizon can be estimated through exemplar-based matching of photographs with known horizon [104] or through vanishing point estimation. Also, the horizon and camera height can be estimated by detecting objects in the scene that have known height or known probability distributions of height. The ground plane provides a reference frame that is more useful than 2D image coordinates for reasoning about object position and interaction in the scene, but it does not provide much information about the layout of the scene itself.

3.2.3.2 Billboard Model or Ground Plane with Walls

As a simple extension to the ground plane model, we can add vertical billboards or walls that stick out vertically from the ground, providing a sense of foreground objects and enclosure. Such models are suited to rough 3D reconstructions [10, 99] and navigation [164] in corridors and street scenes. These models can provide good approximations for the spatial layout in many scenes, such as beaches, city streets, fields, empty rooms, and courtyards, but they are not very good for indoor scenes with tables, shelves, and counters, where multiple support surfaces play important roles. More generally, billboard models do not provide good approximations for hilly terrain or the internal layout of crowded scenes. For example, a crowd of hundreds of people is not well modeled by a single billboard, and it is usually not possible to pick out the individuals in order to model them separately.

To model the scene as a ground plane with walls or billboards, we must estimate the parameters of the ground plane and provide a region and orientation for each wall. In indoor scenes, straight lines and their intersections can be used to estimate wall boundaries [133]. In Chapter 5, we discuss an approach for outdoor scenes that categorizes surfaces as ground, vertical, or sky, and fits lines to their boundaries. The distance and orientation of each wall can be modeled with two 2D endpoints. Usually, estimating these endpoints requires seeing where an object contacts the ground, which may not be possible in cluttered scenes. In that case, figure/ground relationships, if known, can be used to bound the depth of objects [105]. As Gupta et al. [84] show, it may help to consider common structures, such as buildings, that compose multiple walls as single units that are proposed and evaluated together.

3.2.3.3 3D Box Model

The 3D box model is an important special case of the ground plus walls model. A 3D box provides a good approximation for the spatial envelope of many indoor scenes, with the floor, walls, and ceiling forming the sides of the box. The box model, illustrated in Figure 3.2, can be parameterized with three orthogonal vanishing points and two opposite corners of the back wall.

Because it has few parameters, the box model can often be estimated even in cluttered rooms, where the wall-floor boundaries are occluded [93]. It also provides a more powerful reference frame than the ground plane, as orientation and position of objects can be defined with respect to the walls. This reference frame can be used to provide context for spatial layout and to improve appearance models with gradient features that account for perspective [94, 132].

In addition to texture and color cues, which can help identify wall and floor surfaces, the orientations of straight line segments and junctions formed by line segments can be used to estimate wall regions and boundaries [93, 133]. In Chapter 5, we'll discuss an approach to model indoor scenes with a 3D box.

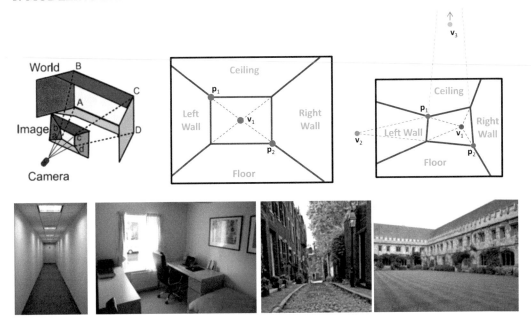

Figure 3.2: Illustration of the 3D box model. Top, left: the projection of the scene onto the image. Top, center: a single-point perspective box, parameterized by the central vanishing point and two corners of the back wall. Top, right: a general box, parameterized by three vanishing points and two corners of the back wall. Bottom: several examples of scenes that are well-modeled by a 3D box (photo credits, left to right: walknboston (Flickr), William Hook (Flickr), Wolfrage (Flickr), Alexei A. Efros).

3.2.3.4 Grammar-based Structures

More flexible structures can be modeled with grammars, or parses that define the shapes and spatial relations of the scene components. Many of the earliest efforts at 3D scene modeling, such as Robert's 1965 "Blocks World" [188] employed grammars. Recently, Gupta et al. adapted this model (shown in Figure 3.1) to incorporate statistical image analysis methods. In Gupta et al.'s model, the scene is represented with a series of 3D blocks and planes and their positional and relationships. Several examples of parses can be found at `http://balaton.graphics.cs.cmu.edu/abhinavg/3D_Parse/index.html`. Han and Zhu [88] provide another good example; they parse a scene into rectangles, cubes, and other line-based structures that reflect the underlying 3D geometry. Other works parse architectural scenes into floors, walls, and windows [124, 248].

3.2.4   LOOSELY STRUCTURED MODELS: 3D POINT CLOUDS AND MESHES

More generally, we can represent the scene geometry as a 3D point cloud or mesh. For example, Saxena et al. [197] estimate depth and orientation values for each pixel. Rather than reducing the parameterization through a fixed structure, these general models achieve generalization through data-driven regularization. In Saxena et al.'s approach, surface orientation are strongly encouraged but not constrained to be vertical or horizontal through terms in the objective function. Likewise, planarity is encouraged through pairwise constraints.

These more general 3D models may be better suited for 3D reconstruction in the long run because they apply to a wide variety of scenes. However, the model precision makes accurate estimation more difficult, leading to frequent artifacts that may reduce visual appeal. Such models are also well suited to robot navigation, which may require more detail than provided by the billboard or box models and offers the possibility of error correction as the robot moves.

The main disadvantage of point cloud and mesh models is that they provide limited abstraction for higher level reasoning. For example, it may be difficult to identify which surfaces can support a coffee mug based on a point cloud.

3.3   SUMMARY

We can think of a 3D scene in terms of its subcomponents (surfaces, objects, materials, boundaries, junctions, light sources, shadows, and physical relations) or geometric structures (e.g., ground plane, walls, depth maps, point clouds). Major tradeoffs in modeling include precision vs. abstraction, viewpoint-centric vs. scene-centric, and holistic vs. modular. For example, 3D point clouds are precise but provide poor abstractions, while the 3D box model provides a good reference frame for spatial reasoning but lacks detail. Viewpoint-centric depth maps and surface orientation maps are easy to align, but the scene representation depends on the camera position. Holistic models, such as the stages of Nedovic [166], can be robustly estimated from an image, but they are not as flexible as the decomposable grammar-based models, such as Blocks World [84, 188].

We consider the application of the discussed models to different types of scenes:

- *Indoor scenes* are highly structured, with orthogonal floors and walls, but are also full of occluding objects, and nearby objects have strong perspective effects. The 3D box model of Hedau et al. [93] attempts to provide the rough room shape in a way that is robust to clutter, but applicability is limited to rectangular spaces. The floor/walls model of Lee et al. [135] is appropriate for more general floor layouts but requires non-occluded floor/wall or floor/ceiling boundaries. Very recently, Gupta et al. [85], used a combination of these models to help infer supporting and resting surfaces, recovering the potential for actions in a room. The stages model [166] is useful for representing canonical scenes depicted on television, such as a news anchor sitting behind a desk.
- *Outdoor urban scenes* are also strongly structured and tend to have highly textured walls and open areas that facilitate geometric estimation. The ground/walls models and many of the grammar-based models are suitable for these scenes. Detection of cars and pedestrians can aid in estimating the ground geometry [101].
- *Outdoor rural scenes* pose the greatest challenge for both single-view and multi-view estimation methods. The scenes typically contain irregular textures, without prominent perspective cues.

In addition, the geometries can vary greatly with hills, groves of trees, cliffs, beaches, and so on. For more open scenes, the techniques of Hoiem et al. [99] and Saxena et al. [200] often work well.

CHAPTER 4

Categorizing Images and Regions

Once we have decided on how to represent the scene, we must recover that representation from images. As discussed in the background (Chapter 1), early attempts at scene understanding involved many hand-tuned rules and heuristics, limiting generalization. We advocate a data-driven approach in which supervised examples are used to learn how image features relate to the scene model parameters.

In this chapter, we start with an overview of the process of categorizing or scoring regions, which is almost always a key step in recovering the 3D scene space from a single image. Then, we present some basic guidelines for segmentation, choice of features, classification, and dataset design. Finally, we survey a broad set of useful image features.

4.1   OVERVIEW OF IMAGE LABELING

Although there are many different scene models, most approaches follow the same basic process for estimating them from images. First, the image is divided into smaller regions. The regions could be a grid of uniformly shaped patches, or they could fit the boundaries in the image. Then, features are computed over each region. Next, a classifier or predictor is applied to the features of each region, yielding scores for the possible labels or predicted depth values. For example, the regions might be categorized into geometric classes or assigned a depth value. Often, a post-processing step is then applied to incorporate global priors, such as that the scene should be made up of a small number of planes.

Many approaches that seem unrelated at first glance are formulated as region classification:

- **Automatic Photo Pop-up** [99]:
    1. Create many overlapping regions.
    2. Compute color, texture, position, and perspective features for each region.
    3. Based on the features, use a classifier to assign a confidence that the region is good (corresponds to one label) and a confidence that the region is part of the ground, a vertical surface, or the sky.
    4. Average over regions to get the confidence for each label at each pixel. Choose largest confidence to assign each pixel to "ground", "vertical", or "sky".

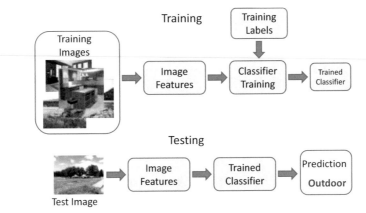

Figure 4.1: Overview of the training and testing process for an image categorizer. For categorizing regions, the same process is used, except that the regions act as the individual examples with features computed over each.

    5. Fit a set of planar billboards to the vertical regions, compute world coordinates and texture map onto the model.

- **Make3D** [198]:
    1. Divide the image into small regions.
    2. Compute texture features for each region.
    3. Predict 3D plane parameters for each region using a linear regressor.
    4. Refine estimates using a Markov Random Field model [143], applying pairwise priors such as that neighboring regions are likely to be connected and co-planar.

- **Box Layout** [93]:
    1. Estimate three orthogonal vanishing points.
    2. Create regions for the walls, floor, and ceiling by sampling pairs of rays from two of the vanishing points.
    3. Compute edge and geometric context features within each region.
    4. Score each candidate (a set of wall, floor, and ceiling regions) using a linear classifier. Choose the highest scoring candidate.

In the above descriptions, critical details of features, models, and classification method were omitted. The point is that many approaches for inferring scene space follow the same pattern: divide into regions, score or categorize them, and assemble into a 3D model.

    Classifiers and predictors are typically learned in a training stage on one set of images and applied in a testing stage to another set of images (Figure 4.1). For region classification, the sets of feature values computed within each image are examples. In training, both labels and examples are provided to the learning algorithm, with the goal of learning a classifier that will correctly predict the label for a new test example. The efficacy of the classifier depends on the how informative the features are, the form and regularization of the classifier, and the number of training examples.

Although there are many possible prediction tasks, the same basic features apply to all of them: color, gradients, histograms, of gradients, edges, histograms of edges, position, and region shape. See Figure 4.2 for examples of several color spaces, gradient features, and edges.

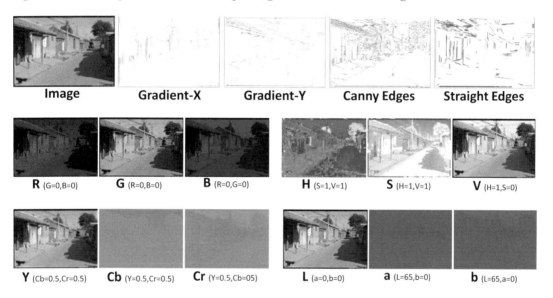

Figure 4.2: Gradients, edges, and examples of color spaces for the input image.

4.2   GUIDING PRINCIPLES

The following are loose principles based on extensive experience in designing features and in application of machine learning to vision.

4.2.1   CREATING REGIONS

When labeling pixels or regions, it is important to consider the spatial support used to compute the features. Many features, such as color and texture histograms must be computed over some region, the spatial support. Even the simplest features, such as average intensity tend to provide more reliable classification when computed over regions, rather than pixels. The region could be created by dividing the image into a grid, by oversegmenting into hundreds of regions, or by attempting to segment the image into a few meaningful regions. Typically, based on much personal experience, oversegmentation works better than dividing the image into arbitrary blocks. Useful oversegmentation methods include the graph-based method of Felzenszwalb and Huttenlocher [65, 100], the mean-shift algorithm [35,

264], recursive normalized cuts [161, 209], and watershed-based methods [3]. Code for all of these is easily found online. Methods based on normalized cuts and watershed tend to be more regular but also may require more regions to avoid merging thin objects. With a single, more aggressive segmentation, the benefit of improved spatial support may be negated by errors in the segmentation. A successful alternative is to create multiple segmentations, either by varying segmentation parameters or by randomly seeding a clustering of smaller regions. Then, pixel labels are assigned by averaging the classifier confidences for the regions that contain the pixel.

Depending on the application, specialized methods for proposing regions may be appropriate. For example, Hedau et al. [93] generates wall regions by sampling rays from orthogonal vanishing points, and Lee et al. [133] propose wall regions using line segments that correspond to the principal directions. Gupta et al. [84] propose whole blocks based on generated regions and geometric context label predictions.

4.2.2 CHOOSING FEATURES

The designer of a feature should carefully consider the desired sensitivity to various aspects of the shape, albedo, shading, and viewpoint. For example, the gradient of intensity is sensitive to changes in shading due to surface orientation but insensitive to average brightness or material albedo. The SIFT descriptor [153] is robust to in-plane orientation, but that discarded information about the dominant orientation may be valuable for object categorization. Section 4.3 discusses many types of features in more detail.

In choosing a set of features, there are three basic principles:

- Coverage: Ensure that all relevant information is captured. For example, if trying to categorize materials in natural scenes, color, texture, object category, scene category, position within the scene, and surface orientation can all be helpful. Coverage is the most important principle because no amount of data or fancy machine learning technique can prevent failure if the appearance model is too poor.

- Concision: Minimize the number of features without sacrificing coverage. With fewer features, it becomes feasible to use more powerful classifiers, such as kernelized SVM or boosted decision trees, which may improve accuracy on both training and test sets. Additionally, for a given classifier, reducing the number of irrelevant or marginally relevant features will improve generalization, reducing the margin between training and test performance.

- Directness: Design features that are independently predictive, which will lead to a simpler decision boundary, improving generalization.

4.2.3 CLASSIFIERS

Features and classifiers should be chosen jointly. The main considerations are the hypothesis space of the classifier, the type of regularization, the amount of training data, and computational efficiency.

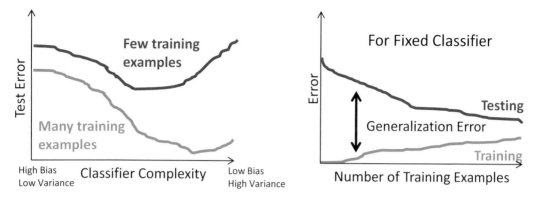

Figure 4.3: Left: As the complexity of the classifier increases, it becomes harder to correctly estimate the parameters. With few training examples, a lower complexity classifier (e.g., a linear classifier) may outperform. If more training examples are added, it may improve performance to increase the classifier complexity. A highly constrained (low complexity) classifier will have low "variance", meaning that similar parameters will be learned from different random sets of examples, but may have "bias", which is the error that results when the hypothesis space is so constrained that the true function cannot be learned. **Right**: As more training examples are added, it becomes harder to fit them all, so training error tends to go up. But the training examples provide a better expectation of what will be seen during testing, so test error goes down.

See Figure 4.3 for an illustration of two important principles relating to complexity, training size, and generalization error.

In the discussion below, we will refer to the logistic regression classifier, which learns a weighting of the features that predicts whether an example should be assigned a positive or negative label. The weights $\mathbf{w}$ are learned to maximize the likelihood of the labels $\mathbf{y}$ ($y_i \in \{-1, 1\}$) given the features $\mathbf{x}$ for the training data, while preferring small weights:

$$\mathbf{w}_{best} = \underset{\mathbf{w}}{\operatorname{argmax}} \sum_{i=1}^{N} \log \mathrm{P}\left(y_i | \mathbf{x}_i; \mathbf{w}\right) - \lambda \|\mathbf{w}\|^2 \tag{4.1}$$

$$\mathrm{P}\left(y_i | \mathbf{x}_i; \mathbf{w}\right) = \frac{1}{1 + \exp\left(-y_i \mathbf{w}^\mathsf{T} \mathbf{x}_i\right)} . \tag{4.2}$$

Specifically, this function maximizes the total log likelihood of the labels of the training set conditioned on the corresponding features, minus a regularization term proportional to the norm of the weights, assuming that examples are independently sampled from the same distribution. The parameter λ determines the tradeoff between maximizing the label likelihood and minimizing the weight magnitudes.

The **hypothesis space** is determined by the form of the decision boundary and indicates the set of possible decision functions that can be learned by the classifier. Sometimes the hypothesis space is measured by the VC-dimension, which is the maximum number of examples for which a classifier could be parameterized to correctly predict the label, for any possible labeling. For example, the linear logistic regression classifier described above has a VC dimension that is $N_f + 1$, where N_f is the number of features. As another example, consider the nearest neighbor classifier which outputs the label of the training example whose features are most similar to the input features. Because the nearest neighbor classifier can accommodate any arbitrary labeling of training examples, it has an infinite VC-dimension.

When there are few examples and many features or parameters, it is likely that some features will coincidentally lead to very confident correct predictions for the training examples but not be predictive for new examples. To avoid such cases, it is common to incorporate a **regularization** term to express a preference for some hypotheses (classifier parameters) over others, typically penalizing more complex decision functions. The regularization term in L2 logistic regression ($-\lambda \|\mathbf{w}\|_2^2$) encodes a preference that no particular feature should have too much weight. Support Vector Machines (SVMs) include a similar regularization term. Sometimes L1 regularization is used instead ($-\lambda \|\mathbf{w}\|_1$), which specifies that the total absolute value of weights should be as small as possible, typically leading to a small set of non-zero weights.

If few training examples are available, then a classifier with a simple decision boundary (small hypothesis space) or strong regularization should be used to avoid overfitting. Overfitting is when the margin between training and test error increases faster than the training error decreases. If many training examples are available, a more powerful classifier may be appropriate. Similarly, if the number of features is very large, a simpler classifier is likely to perform best. For example, if classifying a region using color and texture histograms with thousands of individually weak features, a linear SVM is a good choice. If a small number of carefully designed features are used, then a more flexible boosted decision tree classifier may outperform.

Although out of scope for this document, it is worthwhile to study the generalization bounds of the various classifiers. Generalization bounds are not good for predicting performance on a particular problem, but the bounds do provide insight into expected behavior of the classifier with irrelevant features or small amounts of data.

As a basic toolkit, we suggest developing familiarity with the following classifiers: SVM [205] (linear and kernelized), Adaboost [75] (particularly with decision tree weak learners), logistic regression [167], and nearest neighbor [46].

4.2.4   DATASETS

See Berg et al. [15] for a lengthy discussion on datasets and annotation. In Chapter 8, we introduce several datasets for 3D object recognition. The main considerations in designing a dataset (assuming that a representation has already been decided) is the level of annotation, the number of training and test images, and the difficulty and diversity of scenes. More detailed annotation is generally better, as parts of the annotation can always be ignored, but cost of collection must be considered. More data makes it easier to use larger feature sets and more powerful classifiers to achieve higher performance.

The issue of **bias** in datasets must be treated carefully. Formally, bias occurs when the training and test examples are drawn from different distributions. Informally, a dataset is often called biased if its images are not representative of typical scenes or photographs. All datasets have some bias, in the informal sense. Bias could be due to the acquisition or sampling procedure, to conventions in photography, or social norms. For example, a random selection of photos from Flickr would have many more pictures of people and many more scenes from mountain tops than if you took pictures from random locations and orientations around the world. Bias is not always bad. If we care about making algorithms that work well in consumer photographs, we may want to take advantage of the bias, avoiding the need to achieve good results in photographs that are pointed directly at the ground or into the corner of a wall. The structure of our visual world comes from both physical laws and convention, and it would be silly not to take advantage of it.

But we should be careful to distinguish between making improvements by better fitting the foibles of a particular dataset or evaluation measure and improvements that are likely to apply to many datasets. As a simple example, the position of a pixel is a good predictor of its object category in the MSRC dataset [211], so that including it as a feature will greatly improve performance. However, that classifier will not perform well on other datasets, such as LabelMe [192], because the biases in photography are different. Likewise, it may be possible to greatly improve results in the PASCAL challenge [53] by improving the localization of bounding boxes, but that improvement may not apply to other evaluation criteria.

Although it may seem that big datasets avoid bias, they don't. Whether you sample one hundred or one hundred million examples from Flickr, a large number of them will be framed around people looking at the camera, which is different than what would be observed on an autonomous vehicle. In fact, big datasets make it easier to exploit the foibles of the dataset by making large feature sets more effective, so that care with dataset bias is more important than ever. See Efros and Torralba [48] for an interesting discussion of dataset bias.

In summary, we cannot avoid bias, but we should consider the biases that are encoded by a particular dataset and consider whether measured improvements indicate a better fit to a dataset or a more general phenomenon. Towards this, we encourage experimentation in which training and test sets are drawn from different sources.

4.3 IMAGE FEATURES

4.3.1 COLOR

Color is predictive of material and lighting and is helpful for segmenting regions into separate objects and materials. By far, the greatest variation in color is due to changes in the luminance or brightness. For this reason, color spaces that decouple luminance, such as YCbCr, HSV, and CIELAB, may be preferable to RGB. **YCbCr** is quickly computed and commonly used in display electronics, such as televisions. **HSV**, which decomposes into hue, saturation, and value, is the most intuitive, although the angular measurement of hue (e.g., the maximum and minimum values are both red) can be a nuisance. **CIELAB** is designed to be perceptually uniform, meaning that small changes in color with equal Euclidean distance will be perceived by humans as having similar degrees of change.

Luminance encodes most of the useful information for scene understanding. People can understand grayscale images and movies without difficulty. From luminance alone, most changes in albedo are perceivable. Additionally, luminance encodes the shading information that is predictive of physical texture and overall shape. To better analyze shape, intensity values are typically not used directly. Instead, gradients and zero-sum filters, which better encode changes in shading, are used.

For categorization, color is often represented using averages or histograms. For example, the mean values of YCbCr indicate the overall brightness, the blueness, and the redness of a region. **Histograms** are estimates of the probability that a randomly selected pixel will have a particular color value. Histograms are computed by counting the number of times that each value is observed and dividing by the total value. It is possible to compute separate histograms for each channel or for all channels jointly. Commonly, the three color channels are quantized into a smaller number of values before counting them. One method is to divide the color space into bins (or cubes for three channels). For instance, the values with R<0.25, G<0.25, B<0.25 could be assigned the a value of 1, with all other possible RGB values assigned to one of 64 (4x4x4) discrete values. Then, these discrete values are counted. Another quantization method is clustering. Typically, a sampling of color values is clustered with K-means [46] to learn a set of cluster centers. Then, new values are assigned a number corresponding to the closest center. The K-means clustering aims to provide a quantization that minimizes the mean squared error of the reconstruction of the original values. The clustering approach is more computationally expensive but usually provides a better representation than simple binning for a fixed number of quantized values.

For segmentation, the difference of mean values or histograms is often used to measure region similarity. To compute the difference of histograms, the measures histogram intersection (HIN) and chi-squared (χ^2) are often used. HIN is faster to compute. χ^2 has attractive theoretical justification and sometimes leads to better performance. HIN or χ^2 are also often used as kernels for SVM classifiers with histogram features.

The gain a better intuition for color, we encourage students to play around with sample images, visualizing them using separate color channels in various color spaces. To compare effectiveness of quantization approaches, it may also be helpful to visualize reconstructions of the images after quantizing the colors.

4.3.2   TEXTURE

Texture is predictive of material and material properties and is helpful for separating regions into different objects and materials. Texture is typically represented by computing statistics of responses to zero-mean filters, such as oriented edge filters, bar filters, or blob filters. Commonly used filter banks include those by Leung and Malik [140] (LM filter bank; see Figure 4.4) and Varma and Zisserman [249] (MR filter banks). Like color, the texture of a region is typically represented by

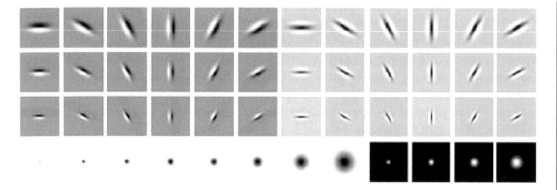

Figure 4.4: Leung-Malik filter bank for representing texture.

computing the average magnitude of filter responses within the region or by computing histograms of quantized values of the filter responses. Histograms of quantized filter responses are often called "textons" [140].

4.3.3 GRADIENT-BASED

Local differences in intensity or gradients are predictive of shape, due to shading, and boundaries of objects and materials. Gradients may computed simply, with filters such as $[-1 \quad 1]$ or $[-1 \quad 0 \quad 1]$ for a horizontal gradient. Often, particularly for detecting important boundaries, the image is first filtered with a Gaussian filter. Gradients can be computed at multiple scales and orientations by changing the bandwidth of the Gaussian pre-filter and by taking weighted averages of the vertical and horizontal filter responses (see the classic work on steerable filters by Freeman and Adelson [73] for details). The magnitude of the gradient between two regions is a strong indicator for the likelihood of an object or material boundary, a depth discontinuity, or a surface orientation discontinuity.

Gradients are also the basis of the most powerful features in object recognition. Oriented gradient values are used to compute many frequently used patch descriptors, such as SIFT [153] and HOG [40] (histogram of gradient). Most descriptors are computed by dividing a patch into different spatial bins (called cells), computing histograms within each spatial bin, and normalizing the values of the histograms. In SIFT, the descriptor is first registered to the dominant orientation so that an in-plane rotation has little effect, and normalization is achieved within each cell by division of total magnitude and thresholding. The HOG descriptor is sensitive to rotation (which is often beneficial for object categorization), and the values in each HOG cell are normalized several times based on the values in the surrounding cells.

HOG features were originally proposed for pedestrian detection [40] and have since been shown effective for other object categories [66] and for a variety of other tasks, such as categorizing scenes and predicting the 3D orientation of materials. HOG features are insensitive to changes in brightness of the light source (due to use of the gradient) and small shifts in position, due to computation of histograms within spatial bins. They are also insensitive to overall changes in contrast, due to the normalization. HOG features are sensitive to shading patterns caused by surface shape and to changes in albedo, and they are sensitive to oriented textures which can predict surface orientation [241]. Object parts, such as eyes, vertical legs, and wheels, often have distinctive HOG feature values. For object categorization, HOG features are often used directly at several fixed positions within a proposed object window. For other tasks, such as scene categorization, histograms of SIFT or HOG features are often computed, in the same way as with color and texture.

4.3.4   INTEREST POINTS AND BAG OF WORDS

The idea of detecting distinctive points was first developed for tracking [90, 236] and then for more general image matching [203] and object instance recognition [152]. Now, interest points and their descriptors, such as SIFT, are used as a general purpose region and image descriptor. After detecting interest points, or placing them at dense intervals along a grid [171], descriptors are computed and quantized. Then, histograms are computed for regions of interest or the image as a whole, depending on the task. Such histogram representations are called *bag of words* models. Many, many papers have been written proposing variants on the descriptor, how many visual words to use, what points to sample [171], faster clustering techniques [169], discriminative quantization techniques, and more. In most cases, simple k-means clustering of densely sampled SIFT or HOG descriptors will work well.

4.3.5   IMAGE POSITION

Due to conventions of photography and the structure of our natural world, image position is often strongly predictive of surface orientation [100], material [211], and object category. Position can be defined within the image coordinates, or relative to an estimated horizon. Especially if defined in image coordinates, position may increase accuracy on the dataset used for training but lead to poorer generalization on other image collections. The position of a region can be encoded by a centroid, a bounding box, or percentiles of sorted positions (e.g., the 5th and 95th percentile of sorted vertical positions for a more robust top and bottom estimate).

4.3.6   REGION SHAPE

Defying intuition and many efforts to devise good representations of shape, region shape tends to provide only a weak cue for categorization in natural images. In part, this is because it is difficult to obtain precise segmentations of objects and surfaces. Also, the 2D silhouette of an object often looks like an indistinct blob. Consider the silhouette of a duck or of a car. These shapes have iconic profiles, but if seen from the front or a three-quarters view, the silhouette will be indistinctive. One simple approach to shape-based categorization is to feed a binary mask into a classifier [187]. Another possibility is to characterize the shape with orientation, eccentricity, major axis length, minor axis length, and so on (see Matlab documentation for `regionprops` for a nice list of simple shape measures).

4.3.7   PERSPECTIVE

Useful perspective cues include statistics of intersecting lines and detected vanishing points, histograms of edges that are oriented consistently with the vanishing points, and more general histograms of oriented straight edges [93, 102, 133]. Such features can provide powerful cues for surface orientation, although histograms of oriented gradient filters can often work nearly as well.

4.4   SUMMARY

Appearance models learned from training images can help to predict geometry for new scenes. One key challenge is to define image feature representations that are robust to unimportant variations but sensitive to important differences. It is also necessary to choose a learning method to specify the parameters of the mapping from the features to the predicted labels. The variety of features and classifiers can be overwhelming, but, in the experience of the first author, using a diverse set of features with a boosted decision tree (if few features) or a linear SVM (if many features) tends to work well. Large improvements typically require careful thought about feature representation or what should be actually be predicted.

CHAPTER 5

Examples of 3D Scene Interpretation

In this chapter, we provide three examples of algorithms for learning and inferring spatial layout from images.

5.1 SURFACE LAYOUT AND AUTOMATIC PHOTO POP-UP

As our first example, we describe a method by Hoiem et al. [99, 102], which first labels the pixels of the image into geometric classes and then constructs a 3D model of the scene based on those labels. The approach uses several techniques and features that are described in Chapter 4. The method is to provide spatial support by clustering small regions into multiple segmentations and using a variety of color, texture, position, and perspective cues to classify the regions into geometric classes. Then, a 3D billboard model (Chapter 3) is estimated by fitting line segments at the boundaries of the bottom of regions classified as "vertical" and regions classified as "support". Because these boundaries are assumed to be on the ground plane, an estimate of the horizon enables the depths of the scene points to be estimated up to a scale (determined by the camera height), using the techniques described in Chapter 2 and assuming a typical focal length.

5.1.1 INTUITION

Humans can easily grasp the overall structure of the scene—sky, ground, relative positions of major landmarks. Moreover, we can imagine reasonably well what this scene would look like from a somewhat different viewpoint, even if we have never been there. This is truly an amazing ability considering that, geometrically speaking, a single 2D image gives rise to an infinite number of possible 3D interpretations! How do we do it?

The answer is that our natural world, despite its incredible richness and complexity, is actually a reasonably structured place. Pieces of solid matter do not usually hang in mid-air but are part of surfaces that are usually smoothly varying. There is a well-defined notion of orientation (provided by gravity). Many structures exhibit high degree of similarity (e.g., texture), and objects of the same class tend to have many similar characteristics (e.g., grass is usually green and can most often be found on the ground). So, while an image offers infinitely many geometrical interpretations, most of them can be discarded because they are extremely unlikely given what we know about our world.

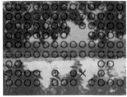

Figure 5.1: Geometric classes. In these images and elsewhere, main class labels are indicated by colors (green=support, red=vertical, blue=sky) and subclass labels are indicated by markings (left/up/right arrows for planar left/center/right, 'O' for porous, 'X' for solid). Figure from [102].

This knowledge, it is currently believed, is acquired through life-long learning, so, in a sense, a lot of what we consider human vision is based on statistics rather than geometry.

The main idea of this approach is to pose the classic problem of geometric reconstruction in terms of statistical learning. Instead of trying to explicitly extract all the required geometric parameters from a single image (a daunting task!), the approach is to rely on other images (the training set) to furnish this information in an implicit way, through recognition. However, unlike most scene recognition approaches which aim to model semantic classes, such as cars, vegetation, roads, or buildings, the goal is to model geometric classes that depend on the orientation of a physical object with relation to the scene. For instance, a piece of plywood lying on the ground and the same piece of plywood propped up by a board have two different geometric classes but the same semantic class. We produce a statistical model of geometric classes from a set of labeled training images and use that model to synthesize a 3D scene given a new photograph.

5.1.2   GEOMETRIC CLASSES

Every region in the image is categorized into one of three **main classes**: "support", "vertical", and "sky". Support surfaces are roughly parallel to the ground and could potentially support an object. Examples include road surfaces, lawns, dirt paths, lakes, and table tops. Vertical surfaces, which are defined as too steep to support an object, include walls, cliffs, curb sides, people, trees, or cows. The sky is the image region corresponding to the open air and clouds.

To provide further geometric detail, we can divide the vertical classes in to several **vertical subclasses**: planar surfaces facing to the "left", "center", or "right" of the viewer, and non-planar surfaces that are either "porous" or "solid". Planar surfaces include building walls, cliff faces, and other vertical surfaces that are roughly planar. Porous surfaces are those which do not have a solid continuous surface. Tree leaves, shrubs, telephone wires, and chain link fences are all examples of porous surfaces. Solid surfaces are non-planar vertical surfaces that do have a solid continuous surface, including automobiles, people, beach balls, and tree trunks.

  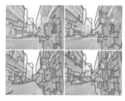 

| Input | Superpixels | Multi Segmentations | Surface Layout |

Figure 5.2: Surface layout estimation algorithm. The input image is oversegmented into superpixels, which are then grouped into multiple segmentations. The final surface layout combines estimates from all of the segmentations. Figure from [102].

5.1.3   APPROACH TO ESTIMATE SURFACE LAYOUT

The algorithm to predict geometric labels from an image is illustrated in Figure 5.2 and outlined below.

1. *Divide the image into hundreds of small regions.* The small regions, called "superpixels", provide spatial support for color and texture histograms. The graph-based method of Felzenszwalb and Huttenlocher [65] is used because it creates relatively few regions (roughly 500 per image) and segments thin objects reliably.

2. *Group the superpixels into multiple larger regions* that provide better spatial support for perspective cues. First, a logistic regressor is trained to predict the likelihood that two superpixels belong to the same surface, based on a comparison of their color, texture, and position. Then, clusters are greedily formed to maximize this pairwise likelihood within the clusters. By varying the number of cluster centers and the initialization, multiple segmentations are produced.

3. *Compute features for each region.* The feature set is designed to encode cues about the material (means and histograms of color and texture), surface orientation (texture, histograms of orientations and intersections of straight line segments, and histograms of edges assigned to vanishing points), and other region properties (size, shape, and position). For color features, the HSV and RGB color spaces are used, and texture responses are measured with the LM filter bank.

4. *Assign the confidence that each region belongs to each geometric class* using boosted decision tree classifiers [34] and the computed features. Also, assign the confidence that each region corresponds to a single surface, using the same features and classification method. The classifiers are learned using regions on training images that have been created using the same multiple segmentation approach and labeled by hand. The boosted decision trees are highly flexible classifiers that output scores based on conjunctions of selected features.

5. *Compute the confidence of each geometric class for each pixel* by averaging over the regions that contain the pixel, weighted by the confidence that the region corresponds to a single surface.

Sometimes, we want to keep the confidence value; other times, we want to choose the label with the most confidence.

5.1.4   EXAMPLES OF PREDICTED SURFACE LAYOUT

On a dataset of 300 outdoor images, using five-fold cross-validation, Hoiem et al. [102] report the average accuracy of the pixel labels to be 88% for the main classes and 62% for the subclasses. Figure 5.3 displays a sample of qualitative results. The same algorithm, when applied to a dataset of indoor images from Delage et al. [42] (replacing "sky" with "ceiling"), achieves an accuracies of 77% and 45% after training on the outdoor images or 93% and 77% when trained on indoor images, for main and subclasses, respectively.

At the time, even the authors were amazed by these results. Theoretically, there is no reason to think that color and texture features would be good predictors for the underlying geometry of a surface. The high accuracy is a testament to the strong regularities of our physical world and of our photographs.

5.1.5   3D RECONSTRUCTION USING THE SURFACE LAYOUT

So far, the geometric labels tell us whether each pixel is part of the ground, an object, or the sky, but it doesn't provide a 3D model by itself. Hoiem et al. [99] take an approach of constructing simple models, like those of a children's pop-up book, by finding where vertical regions contact the ground and folding up at those boundaries.

The photo pop-up algorithm is illustrated in Figure 5.4 and outlined below.

1. *Find connected components for the vertical regions.* These regions will be modeled with a set of connected planes. Sometimes, due to labeling error, regions are loosely connected even though they correspond to distant objects. To improve robustness, a small morphological erosion is used to separate tenuously connected regions before running the connected components algorithm.

2. *Fit a polyline to the base of each vertical region.* Line segments are fit using the Hough transform [45] to the boundary of vertical and support regions. Then, within each region the disjoint line segments are formed into a set of polylines. If no line segment can be found using the Hough transform, a backup procedure is used that greedily searches for a small number of segments that approximates the boundary.

3. *Compute the depth at each point.* Each vertical region with a polyline is treated as a separate object, modeled with a set of planes that are perpendicular to the ground plane. The polyline is on the ground plane. The horizon position is estimated by fitting a horizontal line to the intersecting points of extended line segments (a simplistic but often effective solution). To set camera parameters, skew is set to zero, the aspect ratio to one, and the focal length to a typical value (the length of the image diagonal). Using the techniques discussed in Chapter 2,

Input   Ground Truth   Labels   Input   Ground Truth   Labels

Figure 5.3: Representative sample of results of surface layout algorithm (from [102]), sorted in descending order of main class accuracy. See Figure 5.1 for explanation of colors and markings. Brighter colors indicate higher levels of confidence for the main class labels.

(a) Fitted Segments   (b) Cuts and Folds

Figure 5.4: From the noisy geometric labels, line segments are fit to the ground-vertical label boundary (a) and formed into a set of polylines. Then, the image is "folded" (red solid) along the polylines and "cut" (red dashed) upward at the endpoints of the polylines and at ground-sky and vertical-sky boundaries (b). The polyline fit and the estimated horizon position (yellow dotted) are sufficient to "pop up" the image into a simple 3D model. Figure from [99].

the ground is projected into 3D coordinates. The ground intersection of each vertical plane is determined by its segment in the polyline; its height is specified by the region's maximum height above the segment in the image and the camera parameters. To map the texture onto the model, a separate texture map is created for ground and vertical regions, and the vertical image is feathered. The 3D model is saved in VRML format, which can be viewed on web pages with free plug-ins.

Several successful and unsuccessful attempts at automatic 3D reconstruction are shown in Figures 5.5 and 5.6. Hoiem at al. [99] report a success rate of 30% for input images of outdoor scenes, based on qualitative inspection. Common causes of failure are: 1) labeling error; 2) polyline fitting error, 3) modeling assumptions, 4) occlusion in the image; and 5) poor estimation of the horizon position. The billboard model is too simple to handle crowded scenes, slanted surfaces, and multiple support planes. Also, since no attempt is made to segment connected vertical regions, occluding foreground objects cause fitting errors or are ignored and made part of the ground plane. Additionally, errors in estimating the horizon position (which is quite difficult for many outdoor scenes) can cause angles between connected planes to be overly sharp or too shallow. Overall, though, the ability to automatically create compelling 3D models from one image seemed miraculous, even if it only works on a fraction of images.

Figure 5.5: Good results for automatic single-view 3D reconstruction from the Automatic Photo Pop-up approach. Shown are input images and novel views created from automatically generated 3D models. Figure from [99].

(a)   (b)

(c)   (d)

Figure 5.6: Failures of the Automatic Photo Pop-up algorithm. In (a), some foreground objects are incorrectly labeled as being in the ground plane; others are correctly labeled as vertical but are incorrectly modeled due to overlap of vertical regions in the image. In (b), the reflection of the building is mistaken for being vertical. Major labeling error in (c) results in a poor model. The failure (d) is due to the inability to segment foreground objects such as close-ups of people. Figure from [99].

5.2   MAKE3D: DEPTH FROM AN IMAGE

Make3D [196, 198] is similar to Automatic Photo Pop-up in that it starts by breaking down the image into superpixels and estimates the likelihood that neighboring superpixels correspond to the same surface. However, rather than relying on strong assumptions in the global model, Make3D uses a much more general model and employs local data-driven constraints and priors to regularize the parameters. Rather than assigning regions to one of a few geometric classes, the Make3D algorithm assigns a 3D orientation and depth value to each superpixel. Despite the difference in approach, the behavior of Make3D is often similar to that of Automatic Photo Pop-up. A strong prior is learned that most surfaces are horizontal or vertical, so that with the connectivity priors, the predicted 3D model tends to produce something close to a ground plane with vertical surfaces sticking up from it. Nevertheless, the predictions of Make3D can be more general if sufficient visual evidence is available. Further, the direct prediction of depth makes it easy to incorporate stereo or multiview geometry cues as well, as is shown in [197].

The Make3D algorithm has two parts. The first is to make separate predictions of the depth and orientation of each superpixel. The second is to perform global inference using an MRF model based on local constraints. In training, the parameters for the superpixel predictions and for the local constraints are trained jointly to minimize log difference of predicted depth from ground truth depth. Ground truth is obtained by capturing registered photographs and depth maps from a laser range finder.

5.2.1   PREDICTING DEPTH AND ORIENTATION

The following procedure is used to predict a set of three plane parameters, encoding depth and orientation, for each superpixel:

1. *Divide the image into small regions* (superpixels) using [65].

2. *Compute features for each superpixel*, including the position, color, mean response magnitude and kurtosis of texture and edge filters, and region shape and position.

3. *Predict 3D plane parameters for each superpixel* using a linear predictor. The predictor uses a concatenation of the features within the superpixel and features within four of its neighbors. To account for expected scene variation with changes in position, without resorting to non-linear predictors, the image is divided into 11 horizontal strips, and separate parameters are learned and applied to each.

4. *Compute the confidence in the prediction* using linear logistic regression with the same features as in step 3. It could be learned, for example, that depth predictions for smooth regions are unreliable so that such regions rely more on the pairwise constraints with their neighbors than on their own features.

5. *Perform global inference, incorporating local constraints,* as described next.

5.2.2   LOCAL CONSTRAINTS AND PRIORS

In the natural world, most 3D surfaces are smooth, and many are planar. Make3D takes advantage of this structure by placing a data-driven prior on the plane parameters of pairs of neighboring superpixels. In MRF inference, the depth is predicted that is most likely according to both the individual and pairwise predictions.

The pairwise terms encode the following intuitions (illustrated in Figure 5.7):

• **Connected Structure**: The surfaces corresponding to neighboring superpixels are likely to be connected. For each pair of neighboring superpixels, a linear logistic classifier is used with color and texture features to predict the likelihood of a depth discontinuity between them. A cost is assigned for predicting a depth discontinuity. This cost grows with the difference in predicted depth; the cost shrinks if a depth discontinuity is highly likely based on the features.

• **Co-planarity**: The surfaces corresponding to neighboring superpixels are likely to be part of the same plane. Similarly, to connected structure, a cost is assigned for neighboring pairs that are not co-planar. This cost increases with growth of non-planarity and decreases if a depth discontinuity is likely based on the features. Coplanarity is measured by comparing the predicted depth at a given point using plane parameters for each of the neighboring superpixels. If the same depth is predicted, then the surfaces are coplanar, and no cost is assigned.

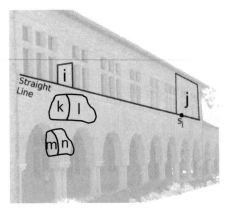

Figure 5.7: Using a Markov Random Field (MRF) model, the Make3D algorithm encode priors that neighboring regions such as (k, l) and (m, n) are likely to be connected. Further, most neighboring superpixels will be co-planar, such as (k, l). Even when not adjacent, superpixels that lie on the same side of the boundary of the same straight line, such as (i, j), are likely to be co-planar. Figure adapted from [198].

- **Co-linearity**: If the same line passes through two superpixels, then the corresponding surfaces are probably co-planar. A cost is assigned for non-coplanarity for superpixels that share a detected line segment, even if the superpixels are not adjacent. The cost increases with increasing non-coplanarity and with the length and straightness of the line that connects the superpixels.

5.2.3   RESULTS

Because the training and test sets contain paired photographs and depth images, the pixel-wise depth predictions of the Make3D algorithm can be evaluated [198]. On average, the $\log_{10}$ error is reported as: 0.300 using only location priors (no image features); 0.205 using only individual superpixel predictions (no pairwise constraints); 0.191 using individual predictions and coplanar constraints; and 0.187 for the full model. These improvements are noticeable quantitatively, but make an even bigger difference qualitatively, where small errors may lead to distracting artifacts. Qualitative results are shown in Figure 5.8 for depth prediction and in Figure 5.9 for texture-mapped 3D reconstructions.

5.3   THE ROOM AS A BOX

In the next example, the goal is to model the visible surfaces of a room and the full extent of the walls. The earlier described methods of Automatic Photo Pop-up and Make3D will not work well for indoor scenes, because rooms are full of objects that obscure wall-floor boundaries.

| Image | Ground Truth | Predicted | Image | Ground Truth | Predicted |

Figure 5.8: Input image, ground truth depth from laser range finder, and predicted depth from Make3D algorithm. Yellow is close, red is medium, and light blue is far. Figure from [197].

The approach of Hedau et al. [93] is to model the scene jointly in terms of a 3D box layout and the surface labels of pixels. The box layout coarsely models the space of the room as if it were empty. The surface labels, similarly to geometric classes from Section 5.1.2, provide pixel localization of the visible object, wall, floor, and ceiling surfaces. To achieve robustness to clutter, three strategies are employed. First, the strongly parameterized 3D box model (see Chapter 3, Section 3.2.3.3) can be estimated robustly from sparse visual evidence. Second, the parameters are estimated jointly using structured prediction based on global perspective cues. Third, clutter is explicitly modeled through the surface labels, so that confusion due to clutter can be avoided when estimating the box. Likewise, the 3D box estimate provides valuable cues to refine the surface label estimates.

5.3.1   ALGORITHM

The algorithm is illustrated in Figure 5.10:

1. *Estimate three mutually orthogonal vanishing points* (Figure 5.10A,B), using a similar edge-based algorithm to that described in Chapter 2, Section 2.4. The vanishing points specify the orientation of the box, providing constraints on its layout.

2. *Generate many candidates for the box layout* (Figure 5.10C) by sampling pairs of rays from two of the vanishing points (Figure 3.2).

3. *Compute perspective cues for each box.* The features are the fraction of edge pixels within each box face that have been assigned (in step 1) to the face's vanishing points.

Figure 5.9: Input image and novel views created using automatic single-view 3D reconstructions from Make3D. Figure from [198].

4. *Compute the confidence of each 3D box hypothesis* using a linear classifier. The classifier is learned using structured learning [242] to minimize the error in predicting the corners of the box and to minimize the overlap of the predicted wall/floor/ceiling regions with the true ones in the training set.

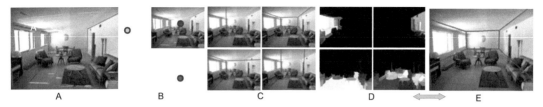

Figure 5.10: This process identifies line segments in the original image (A) and uses them to estimate orthogonal vanishing points (B). The segments in (A) are colored with the color of the vanishing point they vote for. Candidates for the box layout (C) are created by sampling pairs of rays from two vanishing points, and the candidates are ranked based on line membership features. For each highly ranked candidate, surface label confidences are estimated (shown in D for "left wall", "floor", "right wall" and "object"). In turn, the confidence maps are used to compute new features that are used to re-rank the layouts. The final top ranked layout for this image is in (F). Figure from [93].

5. *Estimate the surface labels given the most likely box candidate.* The surface layout algorithm and features (from Section 5.1) are used, with the addition of features that tell what fraction of each region overlaps with the walls, floor, and ceiling of the box layout. Rather than storing the most confident label, confidence maps are stored that indicate the likelihood of each surface label for a pixel (Figure 5.10D).

6. *Re-estimate the box layout using features from the surface labels* (Figure 5.10E). Features are added that indicate the average confidence of each surface label within each box face. Also, to improve robustness to clutter, new perspective features are computed that are weighted by the confidence of object labels (edges within likely object regions have small weight).

5.3.2 RESULTS

On a dataset of 204 training and 104 test images of indoor scenes, Hedau et al. [93] report that the pixel error in surface labels drops from 26.9% to 18.3% when considering the box layout. Additionally, the features from the surface label estimates improve the estimates of the box labels. When measured as the pixel error in the wall/floor/ceiling regions, error drops from 26.5% to 21.2%. When measured as RMS distance between predicted and true corners, error drops from 7.4% to 6.3%, as a ratio of the image diagonal length. Qualitative results are shown in Figure 5.11.

It is possible to get a good sense of the 3D space from the surface labels and box layout, shown in Figure 5.12. Once the camera height and focal length are specified, a 3D model of the walls, floor, and ceiling can be constructed. The three orthogonal vanishing points can be used to recover the focal length and optical center (Chapter 2). The scale is specified by fixing the camera height, and the 3D occupancy of objects is estimated with a heuristic. Using the camera projection matrix, the visual hull corresponding to object pixels is computed (which gives the maximum volume of occupancy). Then, assuming that the objects are supported by floor and are cuboid shaped, the footprint of the

Figure 5.11: A representative sample of results for estimation of the 3D box layout and surface labels [93]. For each image, original image with detected lines is shown in the top row, the detected surface labels in the middle row, and estimated box layout in the bottom row. Lines corresponding to the three vanishing points are shown with red, green and blue color and the outliers are shown in cyan. Each surface label is shown in different color (floor=green; left wall=red; middle wall=yellow; right wall=blue; ceiling=purple; objects=magenta) and the saturation of color is varied according to the confidence of that surface label. Notice that due to the accuracy of estimated vanishing points, most of the images have nearly all correct line-memberships. The estimates of box rotation suffer (e.g., 6th row, 5th column) and the line membership features are not effective (e.g., 6th row, 6th column and 4th row, 2nd column), if the Manhattan world assumption is violated or the available line support in any particular direction is small. These cases account for the highest pixel errors.

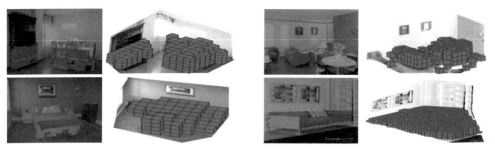

Figure 5.12: 3D model and occupancy computed from the box layout algorithm [93]. The first and third column show the estimated wall boundaries (red lines) and detected object pixels (magenta). The second and fourth column show the 3D model with voxels occupied by objects in magenta. In the two results on the left, the underlying box layout and surface labels are accurate, resulting in a good sense of the 3D space. In the upper-right, the box layout estimate is wrong, and on the lower-right, the surface labels are poorly estimated. Note that small errors in the box layout can lead to considerable errors in the 3D space.

object is estimated by projecting object pixels onto the floor face and used to provide a vertical hull for the object. The intersection of visual hull and vertical hull provides us with the occupied portion of the room. In Figure 5.12, occupied voxels or object pixels are magenta, and the red lines indicate the boundaries of the walls.

5.4   SUMMARY

We have discussed three approaches to estimating 3D scene geometry from an image. The surface layout algorithm of Hoiem et al. [102] labels pixels into one of several geometric classes and then computes a rough 3D model using heuristics based on the ground-object contact points. The Make3D algorithm of Saxena et al. [198] assigns surface orientations and depths to image regions, forming a 3D mesh. Both of these algorithms work best in large-scale scenes, such as open outdoor spaces. The algorithm of Hedau et al. [93] models an indoor room as a simple 3D box taking advantage of the orthogonal structure of indoor scenes and providing robustness to clutter. These methods are just a small sample of recent work, and we encourage the reader to follow up with the references provided in the earlier chapters.

PART II

Recognition of 3D Objects from an Image

In this part of the book, we review methods for recognizing objects from images observed under arbitrary viewing conditions as well as for estimating critical object geometrical attributes. We start with a brief overview of the main theories for representing 3D objects in the human visual system and some of the early computer vision methods in Chapter 6. In Chapter 7, we review recent computational models for 3D object recognition and categorization. We discuss the main characteristics of such models and some of the key learning methodologies in Chapter 8. Finally, in Chapter 9 we present in details a recent state-of-the-art method for 3D object recognition and pose estimation.

CHAPTER 6

Background on 3D Recognition

The ability to recognize objects plays a critical role in scene understanding. When looking at the image of the scene in Figure 6.1, we humans can easily identify all the objects in the scene (e.g., a chair, a desk, etc...), organize them in the scene physical space, and interpret such objects as part of a coherent geometrical and semantically meaningful structure (e.g., the office). How do we do this? It appears that two tasks are critical for accomplishing this goal: i) build representations of 3D objects that allow us to recognize objects regardless of their viewpoint or location in the image. This requires that we have learnt models that are robust with respect to viewpoint changes to enable the identification of object instances or object categories in poses that we have not seen before; ii) infer object geometric attributes such the scale and pose of objects or characteristics about the object shape. These geometrical attributes may provide strong cues for interpreting the interaction among objects in the scene, estimating object functionalities, and, ultimately, for inferring the 3D layout of the scenes as we shall see in more detail in Part III of this book (Figure 6.1c).

6.1 HUMAN VISION THEORIES

The mechanisms that enable the human visual system to categorize 3D objects are yet to be fully understood. It seems easy for us to accurately recognize object instances that we may have (or may not have) seen before regardless of the viewpoint, the background or whether or not these are occluded by other objects. How do we do this? How do we represent 3D objects and perform such a formidable recognition task so effectively? How do we achieve such levels of efficiency and accuracy? How can we scale such ability to as many as thousands of visual categories? Vision scientists have proposed several computational theories that try to explain (with different degrees of success) how objects can be identified within the human visual system. We give an overview of the most important contributions in the next paragraphs. These theories contain several key intuitions that have been subsequently extended and reformulated in more recent computational works for 3D object categorization.

6.1.1 THE GEON THEORY

One of the most influential theories for 3D recognition is the Geon theory introduced by Biederman in the 1980s [17, 18]. The theory introduces the concept of 3D geometric primitives, called *geons*, which are are related to definition of generalized cylinders. The main idea is to describe the object as a graph structure wherein nodes are the geons and edges capture different types of relationships among the geons. Examples of relationship types are side-connected, top-connected, or not connected. By

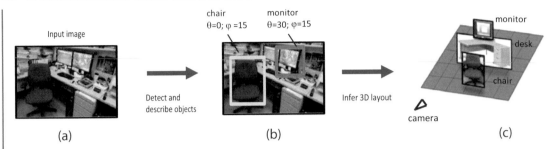

Figure 6.1: We humans can easily identify all objects in the image in panel (a) regardless of their viewpoint or location. This assumes that we have learnt models that enable the identification of objects in poses that we have not necessarily seen before as well as inference geometrical attributes such as pose and scale (b). In the example, the object pose is described by an azimuth angle θ and elevation angle ϕ. The ability to infer object geometrical attributes such the scale and pose helps us organize such objects in a coherent geometrical and semantically meaningful structure in the scene physical space as we shall see in more detail in Part III of this book (c).

combining a set of such elements from an alphabet of geons using different relationship types, it is possible to represent a large number of 3D objects (Figure 6.2). Categorization is carried out by defining a distance metric that can be used to compare the graph structural representation of the query object with that of one of the stored models. The Geon theory has several important properties: i) it introduces an intrinsic 3D representation for the object category whose atomic elements—the geons—are volumetric and, as such, view invariant. This allows a more parsimonious description compared to 2D template based representations as we shall discuss next. Notice, however, that since the relationships among geons are not always view invariant (e.g., the left-side connection), the overall representation is not necessarily view invariant. ii) it enables a hierarchical structural representation of the object wherein the model retains information object parts and their relationship at different level of granularity. These representations are related to early works by [19, 156, 179, 258]. Despite these desirable properties, Geon theory has a number of drawbacks: i) it is unclear how to estimate and recognize the geons from (retinal) images. Although neuronal implementations [113] have been proposed, a plausible method that is also biologically motivated has yet to be introduced. ii) The derivation of structural representations by generalized cylinders is often problematic for certain object categories (think about a tree). iii) the derivation of a structural decomposition is often unstable: the same image of an object can lead to completely different decompositions. Structural representations based on Geons elements have been quite influential among computational methods.

6.1.2   2D-VIEW SPECIFIC TEMPLATES

An alternative and complementary approach to the Geon theory, is the concept of matching 2D-input views to a view-specific representation for an object category. The simplest version of this theory is based on the idea of representing shapes as a single or multiple view-specific templates.

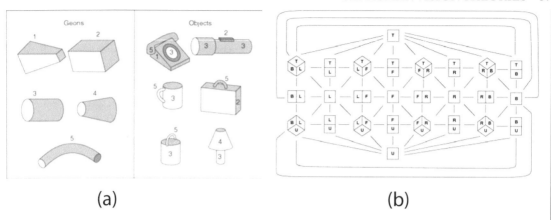

(a)   (b)

Figure 6.2: (a) In the Geon theory an object is described by a graph structure wherein nodes are the geons and edges capture different types of relationships among the geons. Left: Examples of Geons. Right: Examples of objects representations by the Geons). (b) In the Aspect Graph (AG) theory, an object is characterized by a graph structure wherein each node corresponds to an *aspect* of the object, that is a collection of views from which the object topology is constant (i.e., no new facets appear or disappear). Two nodes (aspects) of the graph are connected if at least one object face is shared between the aspects. The figure shows the AG associated to a cube.

Such templates correspond to viewpoints that are more representative than others for humans. This intuition is supported by psychophysical experiments that suggested that some views are recognized faster (with lower naming latency) or more easily than other views for certain object categories. For instance, a horse observed frontally is recognized faster than an horse seen from the top. Human vision scientists have called these views *canonical perspective* [180]. More recently, Tarr and Pinker [227, 228], in a series of experiments, showed that human subjects became faster at recognizing unfamiliar objects from viewpoints that were shown before in a training stage, whereas they were slower if these viewpoints were not shown before. Interestingly, they also showed that the response time in recognizing the object increased as the distance transform between the query viewpoint and closest viewpoint in training increased. The observation that subjects become more accurate and faster at recognizing objects as more viewpoints are presented during the training stage suggests that humans may learn (store in memory) "dictionaries" of canonical view templates for each object category that are used eventually during recognition. While the multiple-view template theory explains convincingly well empirical phenomena such as the naming latency, it suffers from the following limitations: i) it does not naturally provide a structural description of the object in term of parts and their hierarchical dependencies; ii) it requires to store in memory a extremely large number of view-templates in order to effectively capture viewpoint, intra-class and inter-class variability of all existing object categories.

A strategy for lightening the daunting requirement of storing large number of views in template-based methods is suggested by the 2D alignment theories by Poggio and Edelman [184] and Ullman and Basri [247]. In such theories input object views are matched with model views which are either stored in memory (as in standard template-based theories) or *derived* from existing 2D viewpoint templates. For instance, Ullman and Basri [247] show that, under certain restrictions and conditions, it is possible to synthesize a novel view of an object from a linear combination of three 2D views of the same of object. Some of the assumptions are that novel views can be derived only for those regions that are visible from all the interpolating views. Poggio and Edelman [184] show that it is possible to generate novel views from an ensemble of stored views (typically more than 3) using a technique called *generalized radial basis function*. This can be viewed as generalization of the Ullman and Basri [247] theory.

6.1.3   ASPECT GRAPHS

The theory based on aspect graphs, proposed in 1979 by Koenderink and Van Doorn [119], mitigates some of the limitations of the view-specific template representation. The aspect graph theory follows the intuition that nearby views of the same object yield very similar observations. This enables a more parsimonious representation where the number of views that need to be stored as templates can be greatly reduced. The aspect graph (AG) theory assumes that an object can be expressed as a piecewise collection of smooth surface faces (or facets). For instance, a cube has six of such faces. An AG is in essence a graph structure where each node corresponds to an *aspect* of the object, that is a collection of views from which the object topology is constant (i.e., no new facets appear or disappear). Two nodes (aspects) of the graph are connected if at least one object face is shared between the aspects (see Figure 6.2). Important properties of the aspect graph representation are: i) the number of nodes in AG is finite (the graph traversal is discrete) even if the viewpoint changes are continuous; ii) the AG naturally accounts for object self-occlusions: the transition from one node to another of the AG regulates the process of having some object faces to appear or disappear because of self-occlusions; iii) each aspect of the AG can be associated to an object viewpoint template (which effectively captures one specific configuration of object faces) and, thus, the overall number of templates that are required to be stored is greatly reduced. Drawbacks of AG theories are: i) the complexity of the AG representations is a function of the complexity of the object topological structure: objects that are represented by complex piecewise structures would result into AGs with an extremely large number of nodes and connections; moreover the complexity of the AG is function of the level of resolution used to describe the object; the finer the details that are needed to be encoded into the object representation, the more complex is the resulting AG; ii) AGs describe classes of topological structures. For instance, a cube and parallelepiped share the same topology of aspects. Objects that belong to the same class of topological structures cannot be discriminated even if they may be associated to different semantic labels (e.g., an armchair or coach). Notice that this limitation is not suffered by a geons-based or template-based representation wherein quantitative descriptions such as aspect ratios can be easily captured; iii) An AG cannot predict (extrapolate) the appearance of the object from an aspect that is not previously captured in training. iv) Despite the appeal of AG and template-based theories, unlike the Geon theory, representing an 3D object as a collection of (stored or synthesized) 2D views cannot fully account for the sense of three-dimensionality that humans have in perceiving objects.

6.1.4   COMPUTATIONAL THEORIES BY 3D ALIGNMENT

This latter limitation is partially addressed by computational theories that advocate that objects are stored in memory as full 3D models. According to these theories, recognition takes place by matching (aligning) the stored 3D model with the input object view via geometrical transformations (such as rotation, translations and projectivity) [115, 150, 246]. For this reason, these methods are called *pose consistency* or *alignment* theories. The matching procedure has two basic steps: i) hypothesis generation; ii) geometrical verification. This two-step framework has been picked up and extended up in various forms by several computational theories for 3D object recognition in the past three decades. In the hypothesis generation step, a number of correspondences between (a few) features in the input image and features on the 3D model stored in memory are found. These correspondences enable estimation of the viewpoint transformation between observer and the 3D model. In the validation step, all the remaining features characterizing the model are (back) projected into the input image. A score that measures the degree of similarity between the projected features and the observed ones is computed. If the score is above a certain threshold, the hypothesis along with the resulting model is accepted. Because the geometrical verification stage is based on a precise alignment between model and input image, such theory enables a more discriminative recognition procedure than aspect graphs or structural methods such as the Geon theory.

6.1.5   CONCLUSIONS

In summary, structural representations (such as the Geon theory) and template-based (holistic) representations appear to be quite complementary in their strengths and weaknesses. Structural representations are more adequate for representing object categories because of their ability to generalize well and accommodate variations across exemplars within the same category. On the other hand, view-specific methods appear to be more suitable to recognize specific instances of object categories because of their discrimination power. Recently, vision theorists proposed to combine these structural and template-based representations (e.g., Hummel and Stankiewicz [114]) and suggested strategies for using these two models concurrently.

6.2   EARLY COMPUTATIONAL MODELS

The development of theories for 3D object representation in the human visual system is tightly coupled with early and ongoing progress in computer vision research: many of the relevant computational models have been inspired from psychophysical phenomena observed in human subjects. In turn, computational models are used to help predict and validate theories that describe how the human visual system interprets objects and scenes. We already mentioned some of the early approaches. Notable examples are the methods based on pose alignment [60, 61, 115, 151, 154, 191, 247], pose clustering [233], generalized Hough voting [6], geometric hashing [215, 259], random sampling consensus [70] and its variants [237] or by using invariants for indexing recognition hypotheses [71]. The interested reader can find more details in [72]. Such approaches focused on limited class of objects such as rotationally symmetric objects or un-textured polyhedra and found limited applicability to real world objects. Moreover, most of these methods are capable of recognizing single instances of objects (that is, the input object is already seen in training, but under different viewpoints) rather than object categories.

Several computational approaches inspired by aspect graph theories have also been proposed. Aspect Graphs have been defined for simple shapes like polyhedra [210, 216], solids of revolution [49], and curved objects [165]. For details, the reader is encouraged to consult the survey by Bowyer and Dyer [22]. These early approaches suffer the limitation of generating a large number of aspects whose large storage requirement are impractical for any modestly complex object. Eggert et al. [50] observe that the aspect graph is often based on a level of details not fully observable in practice and explore a notion of a scale-space aspect graph to reduce the number of views. Dickinson et al. [44] construct a hierarchical aspect graph system based on a set of primitives. More recently, Cyr and Kimia [39] use the shape similarity metric, extracted from 2D views of the object, to generate a graph structure used in recognition. Similarly, to other computational methods for single instance object recognition, Aspect Graph methods have shown limited success in modeling intra-class variability and have limited ability to handle nuisances such as occlusions and background clutter.

CHAPTER 7

Modeling 3D Objects

In this chapter, we explore models and methods for recognizing 3D object categories from a single 2D image. We concentrate on methods that propose object representations that seek to capture the intrinsic 3D nature of object categories. These methods enable recognition procedures that are robust with respect to viewpoint variability – that is, enable the identification of object instances or object categories in poses that have not been necessarily seen before. Moreover, we focus on methods that allow inference of object geometrical attributes such the scale, pose and shape from a single image. Such object geometrical attributes provide strong cues for interpreting the interaction among objects in the scene and its layout. We review and discuss methods that have been presented to the computer vision community over the past decade.

7.1   OVERVIEW

Recognizing objects from arbitrary viewpoints is a highly challenging problem: not only does one need to cope with traditional nuisances in object categorization methods (objects appearance variability due to intra-class changes, occlusions and lighting conditions), but also to obtain representations that capture the intrinsic viewpoint variability of object categories. In turn, one can use these for detecting objects from images and determining their pose or other geometrical attributes. In the next sections, we review the main models that researchers have proposed over the past decade for addressing these challenges. We start with an overview of methods that aim to address the *single instance 3D object* recognition problem in Section 7.2. In such a problem the object one wishes to recognize is already observed in training, but only from a limited number of viewpoints. The goal is to recognize the object of interest from a potentially novel viewpoint and to determine the object pose. In most of these approaches, objects are represented by highly discriminative and local invariant features related by local geometric constraints. These in turn enable the recovery of object geometrical attributes such as pose and shape. Most of these methods follow the framework explored in early computer vision methods (Chapter 6) of aligning the query image with either a collection of view-templates or a learnt 3D model of the object instance. This process is often carried out using two basic steps in the alignment theories: hypothesis generation and geometric validation (verification) based on geometrical constraints. While critical for ensuring sufficient discrimination power for distinguishing and recognizing single instance objects, geometrical constraints become inadequate in object categorization problems wherein shape and appearance intra-class variability must be accounted for.

In Section 7.3, we review methods that specifically address the object categorization problem. In such problems, the goal is to obtain a suitable representation for object categories that is capable of accounting for appearance and shape variability of object instances within the same object class. Such methods cope with this type of uncertainty by limiting viewpoint variability to either one or very few canonical views (i.e., side-view cars or frontal view faces) following the paradigm introduced in computational theories based on 2D view-point templates (Chapter 6). While endowed with good generalization power, these methods suffer from similar limitations as the 2D view-point templates approaches: they require one to store in memory a large number of view-templates in order to effectively capture viewpoint variability. Moreover, they are not capable of capturing the intrinsic three-dimensional nature of the object category. We call these methods *single views* or *mixture of views* object categorization methods.

Finally, in Section 7.4 and Section 7.5, we introduce a recent paradigm for 3D object categorization where ideas from single-view object categorization and single instance 3D object recognition methods are combined. These methods attempt to address the problem of object classification in a true multi-view setting. We divide these methods in two groups. In Section 7.4, we review *2-1/2D layout models* wherein object diagnostic elements (features, parts, contours) are connected across views to form an unique and coherent 2-1/2D model for the object category (Figure 7.1). Relationships between features or parts capture the way that such elements are transformed as the viewpoint changes. These methods share some key ideas with computational theories based on aspect graphs or 2D or 2-1/2D structural representations (Section 6). In Section 7.5 we review *3D layout models* wherein the object elements are organized in a common 3D reference frame and form a compact 3D representation of the object category (Figure 7.3). Such 3D structures of features (parts, edges) can give rise, for instance, to 3D object skeleton models, 3D star models or hybrid models where features (parts or edges) lie on top 3D object reconstructions or CAD volumes. These methods re-elaborate some of the concepts introduced in the 3D alignment theories and, to a certain degree, the Geon theory (Chapter 6).

7.2   SINGLE INSTANCE 3D CATEGORY MODELS

Beginning from the late 1990s, surfing the wave of important contributions on robust feature detection and description [152, 159], researchers explore novel techniques for addressing the recognition of single 3D object instances. The aim is to move away from the restrictions typically imposed by early methods on the object types (e.g., un-textured polyhedra, symmetric surfaces, no background clutter), but rather consider objects with generic topology, generic appearance properties and observed in cluttered scenes. Single instance 3D object models can be roughly divided into two classes: those that represent objects as collections of 2D view-templates (Section 7.2.1) and those that introduce a global 3D construction for representing the object (Section 7.2.2).

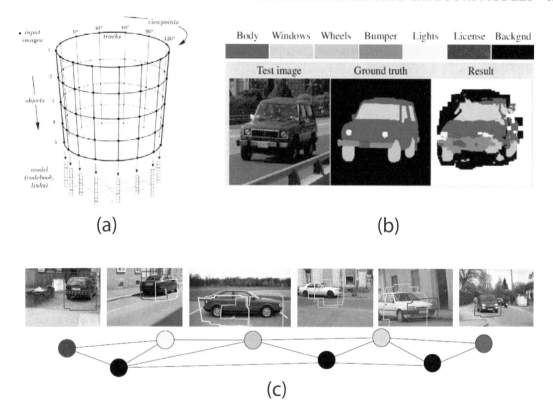

(a)

(b)

(c)

Figure 7.1: (a) In the model by Thomas et al. [232], an object is represented as a 2D collection of features associated to entries of a dictionary of codewords (codebook) for each viewpoint. Codebook entries of each single-view model are connected across views by *links* using [68]. (b) In extension presented in Thomas et al. [231], occurrences of codebook entries are associated with semantic labels describing object regions or other meta-data characteristics. This results into a framework for jointly detecting objects as well as estimating the 3D spatial arrangement (layout) of object semantic regions such as a car wheel or windshield. (c) In the approach proposed in Kushal et al. [127] an object is represented as a loose collection of *pairs* of regions which are linked by local affine transformations.

7.2.1   SINGLE INSTANCE 2D VIEW-TEMPLATE MODELS

Several approaches in the late 1990s and early 2000s explore the idea of using discriminative features and view invariant (e.g., scale, rotation or affine-invariant) descriptors [152, 159, 160] for obtaining object models that are robust to viewpoint variability. Related to the computational theories based on 2D alignment, these methods propose to represent the object as a collection features associated to an handful of 2D view templates. In recognition, by matching features of the query image with those associated to the 2D view templates, it is possible to identify the object of interest. A geometrical verification step is typically applied after feature matching for further validating the quality of the matching and for estimating the pose transformation between the query object and object model. Methods differ by the type of detectors or descriptors or the type of geometric validation process. They have been successfully applied for matching images of objects (or scenes) observed across widely separated views. [32, 152, 158, 172, 185, 201, 202, 229, 245, 267].

A notable example of single instance 2D view-template model is the work by Lowe [152]. Lowe's approach leverages the discrimination power of the SIFT features [152] and relies on identifying SIFT features on a query image that are in correct geometrical correspondences to features associated to training images. Features are first matched to the model by appearance similarity using a generalized Hough transform formulation. This process greatly benefits from the ability of the SIFT descriptor to return a rotationally and scale invariant representation of the image around the point of interest. Then a detailed geometrical verification step is performed based on least square estimation of the object pose via affine or perspective transformation. The method was validated on several objects with moderate viewpoint variation in cluttered scenes (Figure 7.2a).

In order to further increase the robustness of the feature matching process with respect to view transformations, methods have been proposed to group together features into regions that form a consistent geometrical construction such as a 2D affitinity or 2D homography. In Lazebnik et al. [131], groups of local affine regions are identified. These groups are subject to an affine mapping across viewpoint transformations and capture intra-class variations across different instances within the same object class. A validation procedure against a training set is also used to further increase the discrimination power of such regions. Experimental results indicate that discriminative local affine regions can be successfully used to match objects observed from different viewpoints especially in presence of repeated structures (texture) and in images containing significant clutter and noise. Geometrically consistent regions are also studied in the work by Ferrari et al. [68, 69] wherein the concept of group of aggregated matches (GAM) is introduced. Model views are connected by a number of feature-tracks that are used to transfer GAMs across views. GAMs are obtained by applying an iterative technique that alternates between an expansion phase and contraction phase which allows estimation of the optimal spatial extent of the GAM region. By capturing GAM region relationships between the query image and multiple model views, the model enables the recognition of 3D single instance objects. The method is qualitatively and quantitatively evaluated against a dataset of textured objects collected by the authors.

The idea of using segmentation methods to guide the process of grouping feature matches is explored by Toshev et al. [240]. The authors propose a framework for segmenting the images into regions by jointly imposing local perceptually coherency within the same image and feature similarity across different viewpoints. These regions are called *co-salient*. The authors apply this framework for matching urban scene elements against a large database of images of places.

(a)

(b)

Figure 7.2: (a) Detection results from Lowe's method [152]. Lowe's approach leverages the discrimination power of the SIFT features [152] and relies on identifying SIFT features on a query image that are in correct geometrical correspondences to features associated to training images. (b) Detection results from the Rothganger et al. method [190]; in [190], an object is represented as a 3D collection of affine-invariant features (right). Recognition is performed by matching the affine-invariant features detected on the query image with those associated to the learnt models. A geometric validation step is followed based on 3D pose alignment akin to early computational methods

7.2.2   SINGLE INSTANCE 3D MODELS

The object representation in the methods discussed so far is essentially a collection of features or 2D regions associated to a number of template viewpoints rather than to a full 3D object model. The work by Jacobs and Barsi [116] presents one of the early computational models that is built upon the arrangement of regions in the 3D physical space. The method relaxes some of assumptions that previous approaches make on the topology or on the properties of the objects of interest (e.g., rotationally symmetry, un-textured polyhedra). The authors provide a set of provably correct and approximated algorithms based on linear programming for handling the challenging case of self-occlusions and arbitrary occlusions (a problem that, up to date, researchers have yet to find a satisfying answer for). The algorithms were tested on a few single instance textured objects with arbitrary topology such soda cans or boxes and show successful examples of object pose determination.

Rothganger et al. [190] go back to the idea of modeling the 3D object as a collection of points in 3D. The key novelty here is to associate each of the 3D points to an affine-invariant local descriptor. These capture the normalized local appearance of the object in a neighborhood of the image around detected salient features [159]. Recognition is performed by matching features detected on the query image with those associated to the learnt models. Matched features along with global geometric constraints are used to obtain proposals for the object pose which are further validated using a RANSAC-like algorithm [70]. The method was quantitatively evaluated on a database comprising several objects observed under arbitrary viewpoints, in highly cluttered scenes and under severe occlusions. This is one of the largest databases for systematically testing detection accuracy in single instance recognition problems. Anecdotal results on pose estimation were also presented (Figure 7.2b).

The approaches by Brown and Lowe [26] and Gordon and Lowe [77] also represent objects as 3D point clouds and associate each 3D element to a SIFT descriptor. In [77] SIFT features from the input image are matched against each object model. Random subsets of points are chosen and used to estimate a pose hypothesis via RANSAC [70]. The pose is eventually determined by minimizing the re-projection error using optimization techniques. Brown and Lowe [26] focus on the ability to match images of the same object (scene) observed from different viewpoints so to acquire the 3D object (scene) models in a unsupervised fashion from a unordered collection of images containing outliers. This work created the foundation for large scale 3D reconstruction engines such as *Photosynth* [212].

In a recent work, Collet et al. [33] extend these ideas and propose a method for detecting multiple object instances from a single image (i.e., finding objects in an image containing more than one object instance). Recognition is still performed by matching the descriptors between the query image and the stored models, and by applying a geometrical verification step. The novelty relies on using a combination of RANSAC and a clustering technique such as meanshift [35] to avoid the exponential growth in computational time (due to fact that multiple object instances are present). Interestingly, Hsiao et al. [110] demonstrate there is a trade-off between feature discriminability and recognition accuracy. Indeed, by retaining a certain degree of ambiguity in feature matching is it possible to increase the performance of 3D object recognition systems.

7.3   MIXTURE OF SINGLE-VIEW MODELS

Most of the recent advances in object categorization have focused on modeling the appearance and shape variability of objects viewed from a limited number of poses following the 2D view template paradigm introduced in computational theories (Chapter 6). If similar views in the training set are available, and there is a way of modeling such information, the recognition problem is reduced to matching a new image with the known model. This is the approach taken in a number of existing works where each object class model is either learnt from an unique pose [62, 63, 64, 67, 138,

250, 254], or from a mixture of poses [144, 168, 204, 253, 269]. Schneiderman and Kanade [204] propose to use multiple single-view detectors to detect objects in images. Each of these single-view detectors is specialized to specific object orientations and modeled by learning distributions of wavelets coefficients and their position on the object. Weber et al. [253] leverage the 2D constellation model introduced in [254] and represent 3D objects using a mixture of joint probability density on region appearance and shape learnt for each viewpoint. Both [204] and [253] were tested on a few object categories such as faces or cars. The drawback of mixture model approaches is that different poses of the same object category results in completely independent models, where neither features or regions are shared across views. Mixture models typically determines the object pose by returning the viewpoint label of the winning single-view detector in the mixture. Because each single-view models are independent, these methods are are costly to train and prone to false alarms, if several views need to be encoded. An exception is the work by [238] where an efficient multi-class boosting procedure is introduced to limit the computational overload.

Recently, the work by Gu and Ren [80] proposes to use a representation similar to [63] for achieving accurate pose determination. Appearance information is captured using a bank of single-view models of the 3D object. Novel viewpoints are extrapolated (via regression) from the single-view models to enable the modeling of continuous viewpoint angles. The authors show very competitive results in determining discrete object poses on a state-of-the dataset [194] and present one of the first quantitative evaluation of continuous viewpoint estimation. Estimating accurate object pose is also the focus of the work by Ozuysal et al. [178]. Here, the authors propose a two-stage procedure for object detection and pose determination: In a first stage, candidate bounding-box sizes and viewpoints are selected. A subsequent validation stage is followed wherein a view-specific classifier is applied to validate the hypotheses. Even if the method does not enable continuous viewpoint estimation as in [80], compelling results were reported on a 16-pose classification experiment using a car dataset created by the authors.

7.4 2-1/2D LAYOUT MODELS

In *2-1/2D layout models*, object elements (i.e., features, regions, contours) are linked across views so to form an unique and coherent 2-1/2D model for the object category (Figure 7.1). In 2-1/2D models, links between image features (regions) capture the geometrical transformation of the same object physical feature (part) observed across different viewpoints. Thus, the object is not represented as a 3D entity in the physical 3D space but rather as graph structure where nodes are associated to features (regions) and edges capture the geometrical relationship between features (regions). Interestingly, 2-1/2D layout models can be interpreted as multi-view generalizations of 2D part-based models proposed by [64, 254], for single-view object categorization. Moreover, such models are similar to structural representations such as the Geon theory (Chapter 6) wherein geons are here replaced by object 2D regions. Methods based on 2-1/2D layout models differ by the way features (parts) and the relationships between features (parts) are constructed and learnt. In Section 7.4.1, we discuss methods that propose to link features within the same view or across views using Implicit Shape Model representations [137] and its extensions. In Section 7.4.2, we discuss representations constructed on view-invariant parts and their geometrical relationship across views. In Section 7.4.3, we review a proposal for constructing a hierarchical 2D-1/2D layout model where parts are linked across levels of the hierarchy. Finally, we introduce the concept of discriminative aspects in Section 7.4.4 for capturing discriminative appearance properties that are capable of relating objects across viewpoints.

7.4.1   2-1/2D LAYOUT BY ISM MODELS

In the pioneering contribution by Thomas et al. [232], the implicit shape model (ISM) [137]—a 2D single-view object category model—is combined with the single 3D object instance model in Ferrari et al. [68] (Figure 7.1). The ISM is in essence a probabilistic extension of the generalized Hough voting transform [6, 137]. In the ISM model, an object is represented as a 2D collection of features which can be associated to entries of a dictionary of codewords (codebook). Codeword entries are used to cast probabilistic votes for the centroid location of an object category. An ISM model can be learnt for each object viewpoint and the ensemble of these can form a bank of single-view object detectors for the object category. The main idea of Thomas et al. [232] is to connect codebook entries of each single-view model by enforcing across-view constraints to obtain what the authors call the *activation links* (Figure 7.1). Activation links are provided by the dense multi-view correspondences introduced in [69] (7.2.1). This allows the system to obtain an augmented voting space where each feature cast votes for the object class, scale and centroid depending not only on the specific viewpoint but also on the adjacent views. The algorithm was tested on two classes *car* and *bike*. Each model was learnt using a database comprising several instances of objects, where each object instance was observed from multiple viewpoints. Images from the PASCAL VOC 2005 [47] were used for evaluating the ability of the algorithm to detect objects observed from arbitrary viewpoints. Although the framework enables estimation of object viewpoints, the authors did not report pose estimation results.

Several extensions of the ISM model have been proposed to couple the object detector with the ability to estimate object geometrical attributes such as object pose, the 3D object parts' configurations or the object 3D shape. In Thomas et al. [231], occurrences of codebook entries are associated with semantic labels describing object regions or other meta-data characteristics. This results into a framework for jointly detecting objects as well as estimating the 3D spatial arrangement (layout) of object semantic regions such as a wheelchair seat or a car door (Figure 7.1). In the 3D-Encoded Hough voting technique introduced in [222], an object depth map is explicitly used in training for establishing an unique mapping between codeword entries and the relative scale of the object. In recognition, even when 3D information is not available, the authors demonstrate that such mapping produce more consistent votes in the voting space than other state-of-the-art ISM based models. This property helps reduce the sensitivity of the detection method to false alarms. Moreover, the authors show that such framework can be used to estimate the rough 3D shape of the object from a single image. Another notable extension of ISM is discussed in detail in Section 7.5 [4].

7.4.2   2-1/2D LAYOUT BY VIEW-INVARIANT PARTS

As opposed to the ISM model in [232] where single features are linked across views (via codebook entries), and similarly to methods for single object recognition such as [68, 131], 2-1/2D layout models have been proposed where collections of features are grouped into regions and, in turn, regions are linked across views via geometrical transformations. An early example is the work by Bart et al. [14] where an object is represented by a collection of view-invariant regions (or *extended fragments* as termed by the authors). Extended fragments act as templates for measuring similarity across viewpoints. Extended fragments are *view-invariant* in that they are a collection of image templates of the same object physical region under different viewing conditions. The object model, however, is a loose 2-1/2D representation in that it does not does capture any 2D or across-view relationship of regions. The algorithm was validated against one class dataset (*faces*) and showed promising results in recognizing objects from viewing directions that were not seen in the past. Severe viewpoint changes (up 45°) were tested. Other semi-view invariant representations based on constellation of features are explored in [234].

Moving forward toward a more structured 2-1/2D layout representation, the approach proposed in Kushal et al. [127] follows the idea of representing the object as a loose collection of *pairs* of regions (called *partial surface models* or PSMs) which are linked by local affine transformations (Figure 7.1). Unlike Bart et al. [14], however, PSMs are not view-invariant. Similarly, to [131], PSM are local rigid groups of features and are learnt by matching patterns of features across training images within the same category. The affine transformation captures the local geometric relationship of features as the same object part is observed from different viewpoints in the training set. PSM and their relationships are combined into a probabilistic graphical model that effectively captures multi-view appearance and geometry information of the object category. The method showed competitive detection results on the Pascal VOC 2005 [47], but no experiments on pose estimation were reported.

In Savarese and Fei-Fei [194, 195], compact models of an object category are obtained by linking together diagnostic regions (called *canonical parts*) of the objects with the same view or across different viewpoints via affine or homographic transformations (Figure 9.1). Similar to [14], canonical parts are discriminative and view-invariant representations of local planar regions attached to the object physical surface. Unlike [14, 127], however, the object is represented as a full linkage structure capturing the 2-1/2D layout of the object category where nodes are canonical parts and edges encode the geometrical transformation of parts with respect the canonical views. The model is further extended into a generative probabilistic framework in [218, 223]. The main contribution in [218, 223] is the capability to synthesize object part configuration, shape and appearance from arbitrary viewpoint locations on the viewing sphere. We will discuss this approach and its extensions in more detail in Chapter 9.

7.4.3   2-1/2D HIERARCHICAL LAYOUT MODELS

While the idea of representing an object as a collection regions has been successfully used in numerous recognition scenarios, the issue of determining the extent and the correct level of granularity for defining a region still remains an open problem. The model presented in Zhu et al. [271] offers an interesting solution for constructing a hierarchical region-based object representation wherein regions are recursively formed by composition from more elementary region elements. The authors call such regions *recursive compositional models* (RCMs). RMCs are shared (linked) across viewpoints as well as across object instances resulting into a hierarchical 2-1/2D part layout model. Regions are mainly constructed upon object boundaries whereas holistic object appearance (e.g., texture or material properties) is used to provide view-invariant evidence describing the object category. As opposed to [127, 194], however, the model does not capture the geometrical transformation of regions across views, but only the child-parent dependency which is, in essence, a 2D translational relationship for each given viewpoint. The authors quantitatively validated the model's ability to detect objects, determine the object pose and estimate the part layout on an handful of categories from the PASCAL 2005 dataset [47].

7.4.4   2-1/2D LAYOUT BY DISCRIMINATIVE ASPECTS

The methods discussed so far follow the intuition that features and regions can be effectively used to describe geometrical aspects (views) whose combination is used to characterize a 3D object instance or category. An alternative approach is taken by Farhadi et al. [58] where the concept of discriminative aspects is introduced. Discriminative aspects can capture discriminative appearance properties that relate objects across viewpoints regardless of their specific category label or viewpoint label. This allows to capture the relationship between aspects and appearance properties rather than the relationship between appearance and viewpoint label across object classes. Since such discriminative aspects are unknown in training (one of the goals of their method is to discover these aspects automatically), Farhadi et al. [58] propose to represent the discriminative aspects as latent variables using a bilinear latent model similar to [74]. The authors demonstrate that this paradigm has the appealing property of transferring aspects across views and allows to recognize object aspects which are never seen in training.

7.5   3D LAYOUT MODELS

In *3D layout models* the object is represented by a collection of diagnostic object elements organized in a common 3D structure in the physical 3D space. In recognition, a registration step aimed at aligning the model with the query object is typically required. This alignment step is in spirit similar to that introduced in the computational theories for human vision (Chapter 6) or for single instance object recognition methods (Section 7.2). Different techniques use different types of diagnostic elements (e.g., features, regions, edges), or different 3D structures (3D object skeletons, 3D star models, hybrid models where features lie on top 3D object reconstructions or CAD volumes). In Section 7.5.1 we

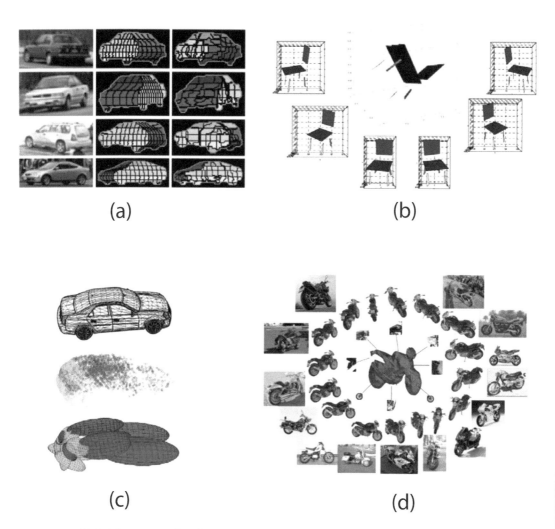

(a)

(b)

(c)

(d)

Figure 7.3: (a) In Hoiem et al. [106], an object category is represented using a 3D layout model which is constructed by grouping together pixels into regions following local and global consistency criteria. The 3D region layout is modeled using a conditional random field (CRF). (b) Chiu et al. [29, 30] propose to represent an object category using the Potemkin model which is a constellation of nearly planar regions localized in a 3D reference system. (c) In Liebelt et. al [145, 146] propose to model the object as a mixture of 3D regions in a 3D common reference frame. Regions are transferred from training images with the help of a CAD model. (d) In Yan et al. [263], the 3D object category is represented as a constellation of 2D features (collected across training images) which are mapped into a 3D reconstruction of the object.

discuss methods that propose representations constructed upon 3D object reconstructions or CAD models (object prototype). In Section 7.5.2 we review methods that propose to represent objects as 3D configuration of features or parts in the object reference system.

7.5.1   3D LAYOUT MODELS CONSTRUCTED UPON 3D PROTOTYPES

In Yan et al. [263] the 3D object category is represented as a constellation of 2D features (collected across training images) which are mapped into a 3D reconstruction of the object (Figure 7.3). Such reconstruction is obtained from a single object instance for which several views are available. The method leverages the 3D shape reconstruction framework proposed in [193]. This allows to automatically establish homographic transformations between the 2D views and the 3D models. This property becomes crucial in detection when hypotheses for the camera pose are to be validated by projecting features of the 3D object model to the query image.

Moving from a feature-based representation to a region-based one, Hoiem et al. [106] propose a 3D layout model which is constructed by grouping together pixels into regions following local and global consistency criteria (Figure 7.3). The former accounts for the coherency of regions within the 2D layout of the object (i.e., 2D relative position of regions should be coherent for each given viewpoint) similarly to [257]. The latter ensures that regions are globally consistent with the position, scale and viewpoint of the object, thus extending the 2D region-based construction into a coherent representation for the 3D object. Global consistency is enforced by introducing a 3D prototype model of the object on top of which training images are matched after suitable alignment. The 3D region layout is modeled using a conditional random field (CRF). The CRF formulation is appealing in enables the object to be identified across viewpoints and also enables part identification and refined object segmentation, as demonstrated in numerous experiments. The authors did not, however, report quantitative results on pose classification.

A part-based representation is also explored in the series of works by Liebelt et al. [145, 146] (Figure 7.3). Unlike [106] or [263] where appearance is associated to a single 3D prototype of the object category, [145, 146] propose to capture 3D shape geometric intraclass variations using multiple CAD 3D models of the object category. Regions are automatically transferred from training images into the CAD models by assuming that images of the objects from several viewpoints are available and that they can be roughly registered with underlying CAD models. This allows to learn mixture of 3D regions in a 3D common reference frame for each object category. Detection follows a two-step algorithm where: i) a object is detected and its pose estimated based on 2D appearance cues; this produces a detection hypothesis. ii) the hypothesis is validated and accurate 3DOF pose is determined using the mixture models via an optimization procedure over the pose parameters. Other ideas based on representing objects via collection of 3D examplars or CAD models are explored in [95, 111, 214, 260]

7.5.2   3D LAYOUT MODELS WITHOUT 3D PROTOTYPES

Unlike [106, 145, 263], the 3D layout representation in Arie-Nachimson et al. [4] is constructed as a 3D constellation of features without the need introducing an underlying 3D prototype. The authors propose to do so by extending the ISM formulation [138] in 3D and by having codebook entries of the ISM to cast probabilistic votes to the object centroid location in a common 3D frame. The authors apply this model for detecting objects and determining its 3D pose using the classic paradigm based on hypothesis generation and geometrical verification: i) determine triplets of matching pairs between features detected in the query images and features in the 3D ISM model; ii) compute a 3D-to-2D similarity transformation using geometrical verifications techniques such as RANSAC. Such transformation allows to accurately estimate the object pose. Quantitative pose estimation results are presented using state-of-the art datasets.

Another representation based on a 3D constellation of regions (without introducing an underlying 3D prototype object model) is proposed in Chiu et al. [29, 30] which the authors call the *Potemkin model* (Figure 7.3). In the Potemkin model, regions are assumed to be nearly planar and are localized in a 3D reference system using the region centroid. Viewpoint transformations are modeled by rigidly transforming the arrangements of region centroids and transforming each region shape as a 2D perspective transformation. The authors propose to use the Potemkin model in two fashions. In [29] the model is used for generating synthetic views of the training objects from a wide range of viewpoints, generating training data for generic multi-view object detection systems. In [30], the authors extend their model to recover geometrical attributes of the object from a single 2D image. These include the 3D region layout, region 3D orientations and auxiliary functional properties. The concept of synthesizing novel views in either training or testing for obtaining more robust object detection or pose estimation is also used in earlier work by Ratan et al. [95, 255] as well as [195, 218].

CHAPTER 8

Recognizing and Understanding 3D Objects

In this chapter, we discuss important practical considerations, including datasets, annotation, and learning strategies, for recognizing and understanding 3D objects.

8.1 DATASETS

Different datasets are used to assess the detection performance of 3D object recognition methods and to evaluate the accuracy in inferring object geometrical attributes such as pose and 3D shape. Traditional datasets used in object categorization [43, 56, 79, 192] are not necessarily suitable to test the ability of 3D object categorization methods to detect objects from arbitrary viewpoints. In such datasets the distribution of viewpoints (under which objects are observed) has strong biases toward certain viewpoints (that is, some of the viewpoints are poorly represented). This, in turn, affects the ability of the algorithms to effectively train and test the models. Moreover, with the exception of the 2006 (and subsequent) releases of the PASCAL dataset (e.g., [54, 55, 56] which assign four viewpoint labels (front, side-left, back, side-right) to some of the object categories, most datasets do not have pose labels. Again, this creates problems for algorithms that require pose annotation in training, or for providing ground truth labels in pose estimation experiments.

One of the earliest dataset specifically designed for training and testing 3D object categorization algorithms is the 3D Object dataset [194]. The 3D Objects dataset contains images of 10 object classes. For each of the object classes, the database comprises images of 10 object instances portrayed from 8 azimuth angles, 3 zenith angles and at three different 3 scales (distances from the observer) for a total of almost 7000 images. Viewpoint labels are also provided. Unlike the PASCAL or LabelMe datasets [54, 56, 192], however, the 3D Objects dataset typically contains only one object instance per image with no occlusions (Fig 8.1).

Other datasets for 3D object categorization and pose estimation have also been introduced. In Thomas et al. [232], images of several instances of motorbikes and shoes are collected from as many as 16 viewpoints. In Xiao et al. [260] images of several instances of motorbikes and shoes are collected from several viewpoints; 3D reconstructions are also provided. The dataset introduced by Lopez-Sastre [149] comprises 26 object categories with (up to) 8 viewpoint annotations per category. In Ozuysal et al. [178] 20 sequences of cars are acquired as they rotate by 360° with a view-angle gap of 3–4° (Fig 8.1). In Zhu et al. [192, 271], a subset of the car dataset from LabelMe [192] is annotated with pose labels. In Arie-Nachimson et al. [4], 86 segmented images of cars seen from

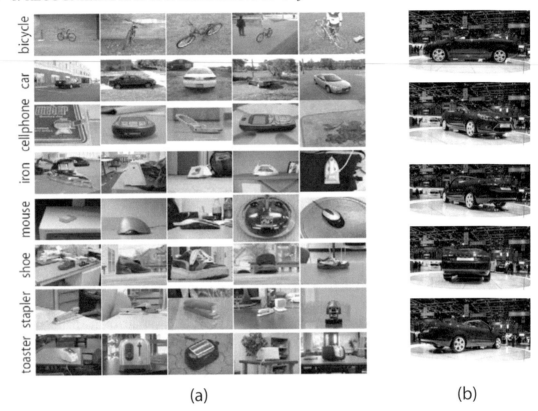

(a)

(b)

Figure 8.1: (a) Images from the 3D object dataset [194]; the dataset includes almost 7,000 images of 10 object categories portrayed from 8 azimuth angles, 3 zenith angles and 3 different scales. (b) The dataset proposed in Ozuysal et al. [178] contains images of 20 *car* instances observed from as many as 90 azimuth viewpoints.

different viewpoints are collected. In Gu and Ren [80], a continuous object pose database is acquired. The database comprises 17 daily objects with a variety of shape, appearance and scale. Images for each object are captured using a turn table while hand-holding a camera and moving it at varying heights and orientations.

A valuable dataset for evaluating the accuracy in estimating region configurations and 3D object layout is presented in [231]. This dataset comprises two classes: wheelchairs and cars. Both classes include ground truth region segmentations and annotations as well as ground truth depth maps. Images of wheelchairs and cars were collected from the Internet and the LabelMe dataset [192]. Finally, a dataset comprising images of three object categories (PC mice, staplers and mugs) observed under a wide range of viewpoints and distributed on a desk top in various configurations is presented in [222]. The interesting property of this dataset is that each image is associated to a depth map. The depth map can be conveniently employed by methods such as [4, 222] that require depth information during training.

8.2   SUPERVISION AND INITIALIZATION

Methods discussed thus far significantly differ by the level of supervision required to learn the 3D models. A typical assumption is that images of objects are labeled according to their object class. Thus, for instance, all images containing car instances are grouped together and used to learn the car category model. Typically, each training image either contains only one object instance occupying a dominant position in the image, or the object is identified by a bounding box.

Some methods require additional level of supervision and necessitate viewpoint labels for either learning or validate the model parameters [4, 80, 106, 127, 223, 232]. While pose annotations facilitate the learning procedure, the process of annotating new viewpoints is problematic in that: i) it requires large amount of human labor for labeling all the views for all of the instances; this is an extremely demanding and laborious task; ii) even by using recent methods such as *Amazon Mechanical Turk* [213], accurate pose annotation is prone to errors as humans are not good at quantifying 3D viewpoints [181].

Techniques such as [14, 194, 195] do not need pose annotations, but require that multiple observations of the same object instance from different views are provided. This is the case when multiple views are required to construct 3D models such as in [263] or to form *canonical* or view invariant regions [14, 194, 195]. This assumption may be problematic if training data are not coherently organized and it is hardly scalable if a large number of categories need to be modeled.

Most of the methods based on 3D layout representations require CAD models or 3D reconstructions as a critical ingredient of the learning procedure. For instance, [106, 263] initialize the learning procedure using a rough 3D reconstruction of a single instance of the object. Moreover, these methods require special care in aligning images to the 3D reconstructions according the viewpoints or need follow some specific guidelines during training. Liebelt et al. [145] assume that several CAD models are available for learning the 3D distribution of regions in a common 3D reference frame. Unlike, [106, 263] however, Liebelt et al. [145] do not require supervision in annotating object regions. Finally, some of the techniques that are capable of inferring 3D region configurations such as [29, 106, 231] require extra supervision in annotating objects regions or in accurately segmenting the object contours in the training set.

Among all the methods for 3D object categorization and pose estimation, Hao et al [218], Zhu et al. [271] and Gu and Ren [80] are those that require the least amount of supervision and allow to learn object models from un-organized and unstructured collection of images of objects portrayed from arbitrary (unlabeled) poses. Both methods also enable the automatic determination of the object regions without the need for extra annotations. Methods such [218], however, do require some degree of supervision for initializing the learning procedure. For instance, the method assumes that one video sequence of a single object instance is available for initializing the model construction. This is required to initialize part correspondence across views. Once this is done, training images can be incrementally added in a unsupervised fashion without the need of extra pose or part annotation. The discriminative approach presented in Gu and Ren [80] do not require any pose labels in training as pose labels are considered as latent variables. Likewise [218], however, model initialization is crucial as the method relies on suitably partitioning the training set.

8.3   MODELING, LEARNING AND INFERENCE STRATEGIES

Methods differ by the modeling strategies, the formulation used to learn model parameters, and the inference procedure. The **probabilistic generalized Hough** voting formulation introduced in Implicit Shape Models (ISM) [137, 232] has been successfully used for learning and capturing dependencies of objects features (via codebook entries) across views. Probabilistic hough voting is also suitable for learning models that allow to infer geometrical attributes such as pose [4, 222], 3D shape [222], 3D part configurations, functional properties and other meta-data associated to object categories [231].

Because of their modeling flexibility, **generative graphical models** are popular tools for modeling 3D object categories. Among the methods based on 2-1/2D layout representations, Kushal et al. [127] propose to use a Markov Random Field to model the relationship among object parts and a loopy belief propagation procedure [163] to approximately learn the parameters of the observed configurations using a validation training set. Another example is the generative graphical model introduced in Su et al. [218], where the process of generating object part shape and appearance is expressed by conditional dependencies on a continuous parameterization of the viewing sphere. Such continuous parametrization is obtained by means of a view morphing formulation that allows to synthesize part location, shape and appearance at any location on the viewing sphere. Parameters of the graphical model are learnt using a EM variational inference method [20]. Geometrical constraints and view-morphing relationships are introduced as penalty terms during variational inference. We discuss this in more detail in Chapter 9. Among the methods based on 3D layout models, Hoiem et al. [106] introduce a conditional random field (CRF) to jointly model the object appearance and the part configuration in 3D. Training leverages randomized trees method [139] to estimate the likelihood of labels given evidence. Inference is carried out by following a optimization algorithm where local consistency of part labeling and layout consistency of the connected components (obtained via message passing [122]) are sequentially enforced. The CRF requires to be initialized using a 3D prototype of the object as already discussed in Section 7.5.

Following the success in 2D object detection problems, discriminative approaches constructed upon latent support vector machines (LSVM) [66] have been also used in 3D object categorization. Gu and Ren [80] use such formulation to achieve high discrimination power in pose estimation problems. In his work, the LSVM formulation, which considers the viewpoint labels as discrete-valued latent variables, has been extended so as to enable continuous viewpoint modeling.

It is interesting to notice that the construction of the building blocks (i.e., features, regions) which are at the foundation of 2-1/2D or 3D layout representations often rely on learning methods. For instance, view invariant and informative object regions are extracted using the measure of mutual information in [14] or multi-model fitting methods such as J-linkage [235] in [194, 195]. Most of the learning and inference strategies used in 3D layout representations leverage geometrical validation schemes such as RANSAC [70] or its variations. The idea is to sample a number of matching candidates between object models and the query object for generating object hypotheses. Such hypotheses are typically obtained by fitting a transformation which allows to align the object model to a certain location and pose in the image. The hypothesis is then validated by verifying that the predicted object location and pose is consistent with the image evidence. Methods such as [4, 33, 77, 190] follow this paradigm.

CHAPTER 9

Examples of 2D 1/2 Layout Models

In this chapter, we review the model proposed in Savarese and Fei-Fei [194, 195] as an illustrative example of a 2D 1/2 layout model. In [194, 195] object categories are modeled as a collection of viewpoint invariant parts connected by relative viewpoint transformations (Sec. 9.1). Moreover, we discuss in detail the extension into the probabilistic framework proposed in [218, 223]. Here, a generative model is used for learning the relative position of parts within each viewpoint, as well as corresponding part locations across viewpoints. The main idea in this work is to introduce a parametrization of the viewing sphere for selecting viewpoints and *generating* novel mixtures of parts (Sec. 9.2).

9.1 LINKAGE STRUCTURE OF CANONICAL PARTS

In the 2-1/2D layout representation introduced by Savarese and Fei-Fei in [194, 195], an object category model is obtained by linking together diagnostic parts (also called canonical parts) of the objects from different viewpoints. Canonical parts are discriminative and view-invariant representations of local planar regions attached to object physical surface. They retain the appearance of a region that is viewed most frontally on the object. For instance, consider a car's rear bumper. This could render different appearances under different geometric transformations as the observer moves around the viewing sphere. The canonical part representation of the car rear bumper is the one that is viewed the most frontally (Figure 9.1a). Similarly, to other part-based methods for recognition, a canonical part is modeled by distributions of vector quantized features [41]. Critically, instead of expressing part relationships by recovering the full 3D geometry of the object as in [26, 68, 190], canonical parts are connected through their mutual homographic transformations and positions (the *linkage structure*). Specifically, given an assortment of canonical parts (e.g., the colored patches in Figure 9.1b), a *linkage structure* connects each pair of canonical parts $\{p_j, p_i\}$ if they can be both visible at the same time (Figure 9.1c). The linkage captures the relative position (represented by the 2×1 vector $\mathbf{t}_{ij}$) and change of pose of a canonical part given the other (represented by a 2×2 homographic transformation $\mathcal{A}_{ij}$). If the two canonical parts share the same pose, then the linkage is simply the translation vector $\mathbf{t}_{ij}$ (since $\mathcal{A}_{ij} = \mathbf{I}$). For example, given that part p_i (left rear light) is canonical, the pose (and appearance) of all connected canonical parts must change according to the transformation imposed by $\mathcal{A}_{ij}$ for $j = 1 \cdots N$, $j \neq i$, where N is the total number of parts

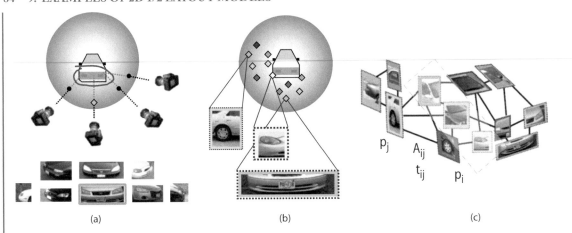

(a) (b) (c)

Figure 9.1: Model Summary. Panel a: A car within the viewing sphere. As the observer moves on the viewing sphere the same part produces different appearances. The location on the viewing sphere where the part is viewed the most frontally gives rise to a canonical part. The appearance of such canonical part is highlighted in green. **Panel b**: Colored markers indicate locations of other canonical parts. **Panel c**: Canonical parts are connected together in a linkage structure. The linkage indicates the relative position and change of pose of a canonical part given the other (if they are both visible at the same time). This change of location and pose is represented by a translation vector and a homographic transformation respectively. The homographic transformation between canonical parts is illustrated by showing that some canonical parts are slanted with respected to others. A collection of canonical parts that share the same view defines a canonical view (for instance, see the canonical parts enclosed in the area highlighted in yellow.

connected to p_i. This transformation is depicted in Figure 9.1c by showing a slanted version of each canonical part.

A *canonical view* V is defined as the collection of canonical parts that share the same view V (Figure 9.1(c)). Thus, each pair of canonical parts $\{p_i, p_j\}$ within V is connected by $A_{ij} = \mathbf{I}$ and a translation vector $\mathbf{t}_{ij}$. A canonical view V can be interpreted as a subset of the overall linkage structure. Effectively, the linkage structure can be interpreted as the generalization to the multi-view case of single 2D constellation or pictorial structure models [64, 67, 254]) wherein parts or features are connected by a mere 2D translational relationship. Similarly, to other methods based on constellations of features or parts, the linkage structure of canonical parts is robust to occlusions and background clutter. The resulting model is a compact summarization of both the appearance and geometrical information of the object categories across views (rather than being just a collection of single-view models).

9.1.1   THE VIEW-MORPHING FORMULATION

An important property of the linkage structure model introduced above is the ability to predict appearance and location of parts that are not necessarily canonical. This capability leverages the view-morphing formulation introduced in the work by Seitz and Dyer [206], and Chen and Willliams [28]. In [28, 206], methods are presented for synthesizing novel views by linear interpolation of two or more existing views of the same scene. This idea is further explored in the view-morphing object model introduced by Sun et al. [218, 223] as we shall see in Sec. 9.2. It is interesting to note that the output of the linkage structure (as well as the view-morphing object model in [218, 223]) is a novel view of the object *category*, not just a novel view of a single object instance, whereas all previous morphing techniques [28, 206] aim at synthesizing novel views of single objects. This is useful to recognize an object observed from arbitrary viewing conditions (that is from views that are not seen in learning) and it is critical for improving false detection rates (a consequence of single-view object representations).

9.1.2   SUPERVISION

Compared to competing techniques based on 2-1/2D or 3D layout representations [106, 127, 147, 232, 260, 263], the construction of a linkage structure of canonical parts requires relatively less supervision. For instance, unlike [106, 127, 147, 232, 260, 263], learning a linkage structure does not require viewpoint annotations and, similarly to [106, 127, 147, 263], it does not require to manually specify the parts needed to build the model. One limitation of the learning procedure, however, is that it does rely on a certain degree of parameter-tuning which prevents the model from achieving good generalization if the intra class variability is high. This limitation, along with the absence of an explicit background model, gives rise to a limited accuracy in object detection and pose estimation. This limitation is specifically addressed in the view morphing models in Sun et al. [218, 223] where model parameters are learnt in a principled way using a probabilistic Bayesian formulation (Sec. 9.2).

9.2   VIEW-MORPHING MODELS

In this section, we give a detailed overview of the 2-1/2D layout representation introduced by Sun et al. [218, 223] which we call the *view-morphing* object model. The view morphing object model has the following main properties: i) **Dense and generative**: If one defines the *unit viewing sphere* as the set of locations from which an object is visible, the view morphing object model is dense and generative in that it has the ability to generate (synthesize) and explain object appearance properties (i.e., features, parts and contours) from every locations on the viewing sphere. ii) **Semi-supervised**: the view morphing object model is semi-supervised in that category labels are provided, but viewpoint labels are not necessarily available. iii) **Incrementally learnt**: The parameters of the view morphing object model are incrementally updated as more unstructured visual data is provided using incremental learning methodologies. Overall, the view-morphing model inherits some of the critical properties from [194, 195] (linkage of parts across views, ability to generate new object views in recognition) while addressing its shortcomings. In the following subsections, we examine details of some of the main properties.

Viewing sphere parametrization and dense representation via view morphing: The viewing sphere is defined as a relatively dense collection of views from which the object is visible from an unit distance (Figure 9.2a). From this collection of views, it is possible to obtain a parameterization of

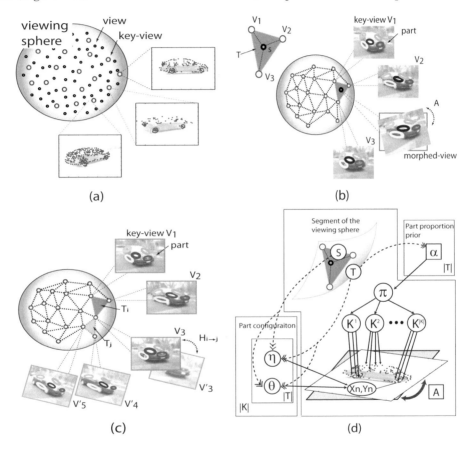

Figure 9.2: Schematic illustration of key concepts of the view morphing model. **(a)** The viewing sphere is defined as the set of locations from which the object is visible from an unit distance. Examples of views from different locations on the viewing sphere are shown. **(b)** The viewing sphere and all the key views are parameterized as a triangle mesh. The figure also illustrates the proposed view morphing machinery. **(c)** Illustration of a 3D geometrical constraint across viewpoint triangles. **(d)** A schematic representation of the part-based generative object model.

the viewing sphere by grouping adjacent triplets of views into triangles T (meshes) and expressing each location within T by a parameter S. S is a 3D vector in the simplex space defined by T. Thus, any position on the viewing sphere can parameterized by $\{T, S\}$ (Figure 9.2b).

Symbol	Meaning
S	morphing parameter
T	triangle on the viewing sphere
A	post-warping transformation
α	part proportion prior
π	part proportion
K	object part
θ	part geometry
m	part center
Σ	part shape
X	feature location
Y	feature appearance

Figure 9.3: Notations used in the view morphing model.

By using the $\{T, S\}$ parametrization of the viewing sphere and some of ideas behind the view morphing theory in [206, 261], one can show that it is possible to to synthesize the appearance of the object in any location within T given the appearance of the object from the three views forming T. Such synthesis is regulated by the parameter S which is defined as the *morphing* parameter. Assume for now that three key views are lying on the same plane (i.e. the views are parallel or rectified). Such a plane can be denoted as a *view plane*. Under the assumption that feature correspondences are available across key views, a new view within T can be synthesized by using the interpolating (*morphing*) parameter S and a homography A, called *post-warping transformation*. The homography A enables the correct alignment (registration) between the synthesized view and a query view. If the assumption of parallel views does not hold (e.g., the key views forming the triangle are not close enough), it is possible to use feature correspondences across the key views to align the key views to the plane formed by the triangle [261]. Note that different triangles may correspond to view planes that are not mutually parallel (Figure 9.2c). In this case, key views can be re-aligned and their viewpoints adjusted through a homographic transformation H (*pre-warping transformation*).

Part-based 3D category generative model over the viewing sphere: For each location on the viewing sphere $\{T, S\}$, the object is represented as a collection of parts. Parts are defined as regions within an object that: i) enclose discriminative features that are frequently observed across different instances of the object class; ii) can be easily identified as the part undergoes viewpoint changes. A part based generative model is learnt following the two criteria above. The generative part-based model is summarized in Figure 9.2c. Following the convention used in probabilistic graphical model representations, each circular node represents a random variable, and each rectangular node represents a parameter. Solid arrows indicate conditional probability relationship between a pair of variables. Dashed arrows indicate the influence of the viewpoint triangle T and morphing parameter S on the variables. The following sets of parameters are introduced for governing the distribution of each object part k under a specific viewpoint: part geometry (θ), part appearance (η) and part proportion (π); π is governed by the Dirichlet parameter α and regulates the likelihood of the different parts that appear under the specific viewpoint. Object part assignments Ks are sampled according to the distribution $Mult(\pi)$. Part geometry θ is in turn expressed as $\theta = [m\ \Sigma]$, where m indicates the part center and $\sigma = [\sigma_x\ \sigma_y]$ describes part shape modeled by its enclosing ellipsoid. Each of these parameters depends on the view triangle T and morphing parameter S. Given m, Σ, the position of each image feature $\hat{X}$ on the view plane is generated according to the Gaussian distribution $\mathcal{N}(m, \Sigma)$. The matrix A is introduced to align (register) the image plane and the view plane. This is also called *post-morphing* transformation. The position of each image feature $\hat{X}$ on the view plane will be equivalent to AX, where X is the position of each image feature on the image plane. Finally, the appearance of each image feature Y is generated from $Mult(\eta)$, where η is the multinomial distribution parameter that governs the proportion of the codewords.

Given the viewpoint parameters, $\{T, S\}$, the view morphing machinery is used to approximately generate part appearance θ and location η parameters from a set of the part parameters $\{\hat{\eta}, \hat{\theta}\}$ of key views in triangle T. Specifically, the part center m is modeled as

$$m = m_T(S) = \sum_{g=1}^{3} \hat{m}_T^g \cdot S^g , \tag{9.1}$$

that is, as the linear interpolation of part centers $\{\hat{m}_T^1, \hat{m}_T^2, \hat{m}_T^3\}$ in the key views of triangle T. Appearance parameter η and part shape parameter Σ can be generated using a similar interpolation scheme.

A key property of this model is that different viewpoints would render very different part distributions (e.g., wheel parts are more likely to be observed from a car's side view than from a frontal view). Such relationship is encoded by the sets of parameters that are governing the part proportion π (with prior α) and distribution of object parts K under a specific viewpoint.

Putting all the observable variables (X, Y, T, S) and latent variables (K, π) together with their corresponding parameters, the joint probability of the proposed model can be expressed as:

$$P(X, Y, T, S, K, \pi) = P(T)P(\pi|\alpha_T)P(S)$$
$$\prod_n^N \{P(x_n|\hat{\theta}, K_n, T, S, A)P(y_n|\hat{\eta}, K_n, T, S, A)P(K_n|\pi)\} . \tag{9.2}$$

A major challenge is to learn the latent variables and estimate the hidden parameters of this model including the morphing parameters such as the pre- and post-warping transformations.

9.2.1   LEARNING THE MODEL

In this section, we give an overview of the training procedure for learning parameters and variables of the view morphing object model. The goal is to enable minimal supervision and incremental update rules as un-organized observations of an object from arbitrary viewpoints become available.

The learning method follows a weakly supervised scheme that is based on critical ingredients such as: i) a suitable initialization seeds the model learning procedure with a meaningful collection

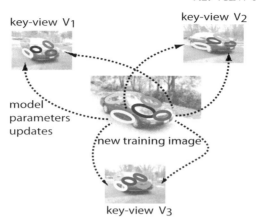

Figure 9.4: Illustration of the updates for position parameter θ during incremental learning. As a new training image is assigned to the triangle T, new evidence on the sufficient statistics is produced which results in updating relevant model parameters of the key-views.

of parts and relationships across views. ii) a number of multi-view geometrical constraints that enables the model learning process to favor those part configurations that have consistent geometrical relationships across different views. As mentioned earlier, weakly-supervised learning methods make the assumption that category labels are provided but that pose labels are not necessarily available. It is also likely that the training data are fairly unstructured. For instance, not all of viewpoints for all object instances are necessarily available (unlike [194]).

Initialization: A promising way for initializing the dense multi-view representation of the 3D object class is by using one or more video sequences of specific object instances. Such video sequences may be acquired by using a hand-held camera and having the camera-person to walk around the object and change viewing angles by raising and lowering the device. The video allows for the obtaining of feature-level correspondences between every consecutive frames. The Lucas-Kanade tracker [155] is employed for estimating such correspondences, and a modified clustering algorithm [235] is used to obtain candidate parts and initialize the part assignment parameter K. There are two advantages for using such video sequence: i) it enables a robust initial condition for parameterizing the viewing sphere and defining its triangulation; ii) it proposes a good initial distribution of parts and part correspondences across views. Both information can be then incrementally updated for accommodating intra-class variability as a larger set of unorganized and unlabeled images becomes available.

Within-T constraint: This constraint enforces that part configurations are consistent across viewpoint transformations within as each triangle T. Specifically, the affine transformation $M_{i \to j}^{T}$ is used to enforce that part centers $\hat{m}_T^i$ and $\hat{m}_T^j$ satisfy $M_{i \to j}^{T} \hat{m}_T^i = \hat{m}_T^j$ for viewpoint V_i and V_j in triangle T, respectively. Tracks (feature correspondences) obtained by Lucas-Kanade algorithm between key views V_i and V_j are used to estimate the affine transformation $M_{i \to j}^{T}$. This information is eventually encoded as a penalty term C in the Variational EM algorithm introduced below. This lets the learning algorithm favor those part configurations that have consistent geometrical relationships across different views.

Across-T constraints: As is shown in Figure 9.2c, each key view is shared by neighboring triangles $\{T_i, T_j, \ldots\}$. This constraint enforces correct correspondences between parts across adjacent triangles and, most importantly, that same parts in different planes (defined by neighboring triangles) share the same configuration in a common coordinate system. Similarly, to the within-T constraint, it is possible to take advantage of the affine transformation $H_{i \to j}$ for relating part centers in the key views belonging to adjacent triangles T_i, T_j. This information too is eventually encoded as a penalty term F in the Variational EM algorithm introduced below. In practice, this constraint serves as a tool for rectifying (parallelize) adjacent views.

Learning the model: The goal of learning is to estimate the hidden variables K, π and part parameters θ, η, α by maximizing the log marginal probability $\ln p(X, Y, T, S)$. The learning problem is formulated as an optimization problem and it is solved by using a modification of the variational EM algorithm [20]. In more detail, the goal is to maximize $\lambda L_q(u) - (1 - \lambda) C(u)$ such that $F(u) = 0$; L_q is the lower bound of log marginal likelihood $P(X, Y, T, S)$; C and F are the penalty terms as defined above; u denotes all the model and variational parameters $\{\hat{\eta}, \hat{\theta}, \gamma, \rho_n\}$, and λ is the weight to balance the importance of the within T constraints C vs. the lower bound L_q. An EM-like procedure can be used for learning model parameter and variational parameter updates, respectively.

9.2.2   DETECTION AND VIEWPOINT CLASSIFICATION

In this section, we describe how to use the view morphing object model to detect and categorize objects in query images as well as recover their pose. The detection process follows these three steps: (1) train robust generic object class detectors; (2) use such detectors to generate hypotheses for the object existence and pose; (3) use the view morphing machinery to implement a model confirmation and pose refinement module.

Step 1: Detectors are implemented for each category for each view by training a random forest classifier [23]. Such classifiers are trained using learnt part model labels and by retaining relative part position (attached to leaf nodes of the random forest) from the object center. Other classifiers (or battery of classifiers) can also be used.

Step 2: Given a test image, the classifiers are used to create hypothesis for object locations, class labels and poses. A random forest classifier is first applied for each object part to obtain candidate locations. Then, part candidates are used to locate candidate object locations by casting votes to the candidate object center using a generalized Hough transform voting scheme [138].

Step 3: This is the most critical step of the detection algorithm. The goal is to employ the view morphing/view synthesis machinery for detection confirmation and for recovering the pose $\{T, S, A\}$. Such parameters are estimated so as to best explain the object pose hypothesized in the query image (top-down model fit for pose refinement). Notice that recovering the pose is no longer a *classification* problem (i.e., find the model among K that best explains the observation), but rather an *estimation* problem on the continuous variables $\{T, S, A\}$. This estimation problem can be formulated as an optimization process where the *argmax* over $\{T, S, A\}$ that minimizes the discrepancy between predicted evidence and observation is computed. After estimating $\{T, S, A\}$, recovering object location, orientation and distance from observer's viewpoint becomes an easy task. T, A allows for the computing of the angular coordinates of the observer on the viewing sphere. A allows for the computing of the distance and in-plane orientation of the object with respect to the observer. Note that, in general, the distance cannot be recovered from a single image. However, using prior information about the *actual* scale of the object category, such distance can be obtained in practice.

9.2.3   RESULTS

The view morphing object model and the associated detection and pose estimation algorithms have been evaluated experimentally in [218] using two datasets: the 3D Object dataset [194] extended with seven household object classes collected from the Internet by the authors, and the car and bicycle classes in PASCAL VOC 2006 dataset [56].

Figure 9.5 compares the ROC detection results of the view morphing object model with other state-of-the-art algorithms on 3DObjects dataset [194]. The view morphing model significantly

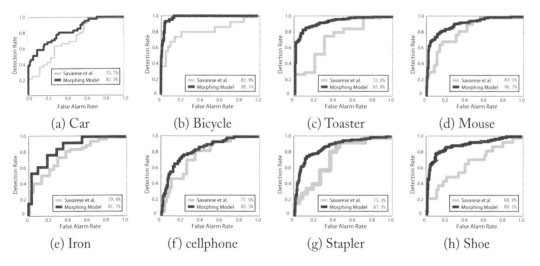

Figure 9.5: Object detection results using the 3DObjects dataset [194] for eight object classes. The ROC curves measure accuracy vs false alarm rate. The area under the curve (AUC) is used to numerically quantify the overall recognition accuracy. Results for the view morphing model are indicated in red and results for Savarese et al. [194] are indicated in green.

outperforms [194] and its probabilistic extension [218, 223] on the 3DObjects dataset. Figure 9.6 compares the precision-recall detection results of our view morphing model with other state-of-the-art algorithms on PASCAL VOC06 dataset [56]. View morphing model shows comparable results

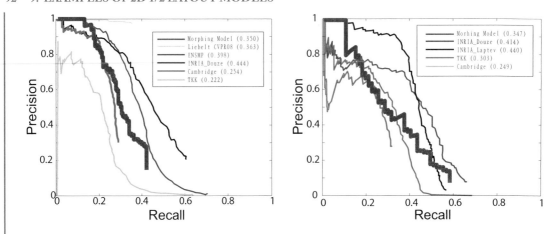

Figure 9.6: Object detection results using the PASCAL VOC06 dataset [56]. **Left** Object detection using the PASCAL VOC06 car dataset. **Right** Object detection using the PASCAL VOC06 bicycle dataset. Precision-recall curves are used to show the results of the view morphing model (red line) compared with [147] and the detection result of the 2006 challenges [56]-INRIA_Douze , [56]-INRIA_Laptev, [56]-TKK, [56]-Cambridge, and [56]-ENSMP. Average precision (AP) scores are shown in the legends.

to most of the state-of-the-art methods on the PASCAL VOC06 dataset. Some example detection results are shown in Figure 9.7.

Further experimental results demonstrate the ability of the model to learn object parts automatically and in turn use them for building an object class detector. Figure 9.8-Left shows that the object detector built by using these proposed object parts significantly outperforms an object detector that does not use the parts. The authors also shows the significance of the dense viewpoint representation and view morphing framework for building 3D object models. Figure 9.8-Center shows a detection experiment on the same car dataset measured by AUC. View morphing model (red curve) is compared with a nearest neighbor model (blue curve). We observe two trends. For both of these models, as the number of viewpoints increases during training, the detection performance increases. The view morphing model, however, performs consistently better than nearest neighbor model even given the same number of viewpoints. This is due to the effect of the morphing parameter S in view morphing model. Such model is capable of synthesizing intermediate views to mitigate the unavoidable discrepancies existing in a discretized representation.

The view morphing model can be used to predict the viewpoint of a query object by estimating $\{T, S, A\}$. Examples of the viewpoint classification results are shown in Figure 9.7. Similar to the object detection experiment, the view morphing model is evaluated against three datasets: the 3DObjects dataset [194], PASCAL VOC06 dataset [56] and the seven classes of Household Objects dataset. The results are shown in Figure 9.9. Note that the numbers of views are provided by the datasets. The 3DObjects includes 8 viewing angles, 3 scales and 2 heights; the PASCAL VOC06

(a) Bicycle   (b) Swivel Chair   (c) microscope

(d) Car   (e) Watch   (f) Iron

(g) Teapot   (h) Flashlight   (i) Calculator

Figure 9.7: Examples of viewpoint estimation for bicycle [56, 194], swivel chair, microscope, car [56, 194], watch, iron, teapot, flashlight, and calculator. Blue arrows indicate the viewpoint T for the detected object (in red bounding box). Green bounding box indicates correct detections of the objects, but in a different viewpoint.

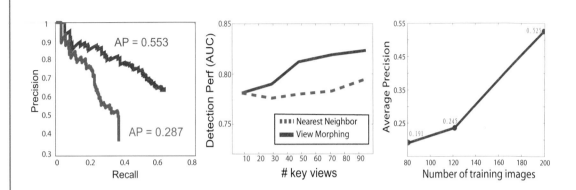

Figure 9.8: **Left:** Object detection with or without using the object parts. A car detection task (3DObjects dataset [194]) is used to show the performance difference between an object detector using the object parts learned by the 3D model (red line) and an object detector built without the object parts (blue line). **Center:** Effect of view synthesis for recognition via learning with the morphing parameter S and number of viewpoints. This effect is demonstrated by showing a binary detection task result (measured by the area under the curve, the AUC) versus the number of key views used by the model. View morphing model (red solid line) is compared with a nearest neighbor model (blue dashed line). **Right:** Effect of incremental learning.

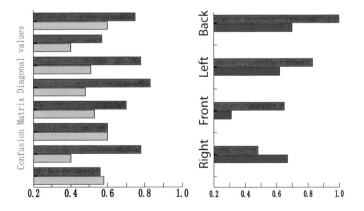

Figure 9.9: Viewpoint classification results: **Left.** 8-view classification of the 3DObjects car dataset. The view morphing model (red bar) is compared with [195] (green bar). **Right.** 4-view classification of the PASCAL VOC06 car dataset. The view morphing model (red bar) is compared with [223] (blue bar).

includes 4 viewpoints. Notice that the view morphing model significantly outperforms [195] on the 3DObjects dataset [194], largely due to its richer representation and view morphing ability's to refine pose estimations.

9.3   CONCLUSIONS

In this chapter, we introduced a general scheme for learning multi-view part-based models of object categories using a generative models and a suitable parametrization of the viewing sphere. These models benefit from the ability to synthesize (generate) the object appearance from novel viewpoints during the recognition stage. Experimental evaluation demonstrated that this is critical for improving the detection accuracy (i.e., the confidence in identifying an object in a certain location can increase once the best object geometrical configuration is determined) and for obtaining a continuous estimation of the object pose. Several important extensions are worth discussing. For instance, similarly to some of the early 3D object representations (e.g., Jacobs and Barsi [116]) or, more recently, to Chiu et al. [30], it is desirable to include into the model the capability to reason about self-occlusions. The goal would be to generate those parts that are observable from the target viewpoint and flag those parts that are no longer visible. Moreover, similar to [106], an additional desirable property is to integrate into the generative view synthesis process the ability to segment and accurately identify the object boundaries for an arbitrary target viewpoint. Both properties may further help increase the confidence of the true positives and decrease that of the false alarms during the detection stage.

PART III

Integrated 3D Scene Interpretation

Scene understanding requires the coordination of many different tasks — occlusion reasoning, surface orientation estimation, object recognition, and scene categorization, among others. In this part of the book, we provide examples from two main strategies for uniting detection of scene geometry and objects into a coherent scene interpretation. The first strategy is to explicitly model and reason about the relationships between the objects and scene geometry. In Chapter 10, we discuss in detail the relations between scene perspective and object size, appearance, 3D pose, and parts layout. We also include a more limited discussion of how to model constraints in the scene layout and the need for ideas to incorporate occlusion reasoning into object recognition. In Chapter 11, we describe a framework for representing scene properties as image-registered maps and using them to compute features that feed back into the individual estimation algorithms.

CHAPTER 10

Reasoning about Objects and Scenes

In this chapter, we discuss how the interaction of objects and scene space can be modeled explicitly through symbolic reasoning or modeling of perspective. Typically, objects, scene categories, and other properties are modeled as variables, and probabilistic relations are defined among these variables. For example, Murphy et al. [162] model the co-occurrence relationships between objects and scene categories. Sudderth et al. [220] infer the depth of points in the image using detected objects. Leibe et al. [136] and Ess et al. [52] model the relationship between objects and scene geometry. Several others (e.g. [83, 142, 186, 266]) model object-object spatial interactions. In the following text, we discuss how perspective, layout, and occlusion relations between objects and scenes can be modeled.

10.1 OBJECTS IN PERSPECTIVE

We can better model the size, appearance, and parts layout of an object by considering the perspective of the camera.

10.1.1 OBJECT SIZE

If we know the camera's perspective and the object's position, we can predict what size the object should be in the image. Likewise, if we can identify objects in the image, we can use knowledge of their likely physical size to estimate the parameters of the camera and a supporting plane. Typically, when first provided an input image, we don't know anything about the camera, the ground plane, or the objects, but we do know how they all relate. In such cases, a probabilistic framework is appropriate.

10.1.1.1 Perspective Projection

Here, we examine the relationship between the camera and object size, which was first introduced in Chapter 2. We assume that all objects of interest rest on the ground plane. This assumption may seem restrictive (cannot find people on the rooftops), but humans often seem to make the same assumption (we fail to notice the security standing on the rooftops at political rallies unless we specifically look for them).

Under this assumption, knowing only the camera height and horizon line, we can estimate a grounded object's height in the scene from its top and bottom position in the image (see Figure 10.1).

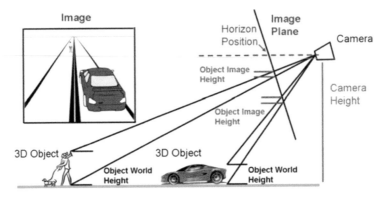

Figure 10.1: An object's height in the image can be determined from its height in the world and the viewpoint. Figure from [101].

We will now derive this relationship using the following notation: pixel coordinates (u, v) ranging from $(0,0)$ at the bottom-left to $(1,1)$ at the top-right; world coordinates (X, Y, Z) with Y being height and Z being depth; camera tilt θ_x; focal length f; camera optical center (u_0, v_0); and camera height Y_c (camera translation $t_y = -Y_c$). By convention, the world coordinates are defined by $t_x = 0$, $t_z = 0$, and the ground plane at $Y = 0$. We assume zero roll (or that the image has been rotated to account for roll) and define the horizon position v_h as the vanishing line of the ground plane in image coordinates. In these coordinates, camera tilt (in radians) is given by $\theta_x = 2 \arctan \frac{v_0 - v_h}{2f}$. We use a perspective projection model with zero skew and unit aspect ratio.

Using homogeneous coordinates, the transformation from image coordinates to scene coordinates is given by

$$
\begin{bmatrix} u \cdot w \\ v \cdot w \\ w \end{bmatrix} = \begin{bmatrix} f & 0 & u_0 \\ 0 & f & v_0 \\ 0 & 0 & 1 \end{bmatrix} \begin{bmatrix} 1 & 0 & 0 & 0 \\ 0 & \cos\theta_x & -\sin\theta_x & -Y_c \\ 0 & \sin\theta_x & \cos\theta_x & 0 \end{bmatrix} \begin{bmatrix} X \\ Y \\ Z \\ 1 \end{bmatrix}.
\tag{10.1}
$$

>From this, we can see that

$$
Y = \frac{Z (f \sin\theta_x - (v_0 - v) \cos\theta_x) + f Y_c}{(v_0 - v) \sin\theta_x + f \cos\theta_x}.
\tag{10.2}
$$

Now suppose that we are given the top and bottom position of an upright object (v_t and v_b, respectively). Letting Y_b be the height at the bottom of the object, $Y_b = 0$ because we assume the object to be on the ground plane. We can solve for object depth Z_b:

$$
Z_b = \frac{-f Y_c}{f \sin\theta_x - (v_0 - v_b) \cos\theta_x}.
\tag{10.3}
$$

>From Equations 10.2 and 10.3, we can solve for object height Y_t:

$$Y_t = \frac{-f Y_c \left(f \sin \theta_x - (v_0 - v_t) \cos \theta_x\right) / \left(f \sin \theta_x - (v_0 - v_b) \cos \theta_x\right) + f Y_c}{(v_0 - v_t) \sin \theta_x + f \cos \theta_x}. \tag{10.4}$$

If the camera tilt is small (e.g., if the horizon position is within the image), we can greatly simplify this equation with the following approximations: $\cos \theta_x \approx 1$, $\sin \theta_x \approx \theta_x$, and $\theta_x \approx \frac{v_0 - v_h}{f}$, yielding:

$$Y_t \approx Y_c \frac{v_t - v_b}{v_h - v_b} / \left(1 + (v_0 - v_h)(v_0 - v_t)/f^2\right). \tag{10.5}$$

In experiments, Hoiem et al. approximate further: $(v_0 - v_h)(v_0 - v_t)/f^2 \approx 0$, giving us

$$Y_t \approx Y_c \frac{v_t - v_b}{v_h - v_b}. \tag{10.6}$$

How valid are these approximations? Equation 10.6 is exact when the camera is parallel to the ground plane (such that $\theta_x = 0$ and $v_h = v_0$). Even when the camera is tilted, the approximation is very good for the following reasons: tilt tends to be small ($v_0 - v_h \approx 0$; $\theta_x \approx 0$); the tops of detected objects (e.g., pedestrians and cars) tend to be near the horizon position since the photograph is often taken by a person standing on the ground; and camera focal length f is usually greater than 1 for the defined coordinates ($f = 1.4$ times image height is typical). However, the approximation may be poor under the following conditions, listed roughly in order of descending importance: object is not resting on the ground; camera tilt is very large (e.g., overhead view); or image taken with a wide-angle lens (f is small). In practice, the approximation is sufficient to improve object detection and to accurately estimate object size in street scenes.

To simplify the notation in the remainder of this section, we will refer to the world height, image height, and bottom position in the image of object i as Y_i, h_i, and v_i, respectively. As before, we denote horizon position v_h and camera height Y_c. Using this notation (illustrated in Figure 10.2), we have the following relationship:

$$Y_i \approx Y_c \frac{h_i}{v_h - v_i}. \tag{10.7}$$

10.1.1.2 Modeling the Scene

We want to determine the camera viewpoint and object identities from the image. Consider the case of car detection. An image patch is likely to be a car if it looks like a car and if it is the right size for a car. We can model the appearance through object detectors. Assuming that each object is on the ground, we can compute its physical height Y_i from its image height h_i and the camera parameters $\theta =\in \{v_h, Y_c\}$. That means if we have a distribution over Y_i, we can compute the likelihood of h_i given θ. So using both object detectors and perspective, we can evaluate the likelihood of a hypothesized object given its appearance and size.

Formally, we are provided a bounding box $\mathbf{b}_i \in \{u_i, v_i, w_i, h_i\}$ in the image, specified by the bottom-center position (u_i, v_i) and the width w_i and height h_i. We use y_i to denote the label of the

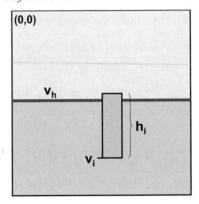

Figure 10.2: Illustration of horizon position v_h, object bottom position v_i, and object image height h_i. With these and camera height Y_c, we can estimate object world height Y_i using $Y_i \approx \frac{h_i Y_c}{v_h - v_i}$. Figure from [104].

bounding box; $y_i = 1$ means that the object exists. The size of each object in the image h_i depends on its size in the world Y_i and on the camera parameters $\theta = \in \{v_h, Y_c\}$. The appearance of an object is scored by a pre-trained detector (e.g., Dalal and Triggs [40]). We consider the detector score d_i as evidence, yielding a likelihood of appearance $P(y_i|d_i)$. We assume that the appearance and size of different objects are mutually independent, when provided the camera parameters. Under this conditional independence assumption, the likelihood of the scene decomposes as:

$$P(\theta, \mathbf{y}, \mathbf{b}, \mathbf{d}) = P(\theta) \prod_i P(\mathbf{b}_i, y_i|\theta) P(d_i|y_i). \tag{10.8}$$

What we really want to know is the likelihood of the camera parameters and the object candidates given hypothesized objects and detector scores, $P(\theta|\mathbf{b}, \mathbf{d})$ and $P(y_i|\mathbf{b}, \mathbf{d})$. To get there, we need to do some probabilistic rearrangement. For simplicity, let's make θ discrete. We assume that, without knowing the label y_i, the bounding box $\mathbf{b}_i$ is independent of θ; likewise, without knowing θ, y_i is independent of $\mathbf{b}_i$. Also, the object bounding box coordinates (u_i, w_i, h_i, v_i) are a priori uniformly likely and independent. Then, $P(\mathbf{b}_i, y_i|\theta) = P(h_i|y_i, v_i, \theta)P(y_i)P(u_i)P(v_i)P(w_i)$. Using Bayes' rule, $P(d_i|y_i) = \frac{P(y_i|d_i)}{P(y_i)}$. Using all of this, we can derive:

$$P(\theta|\mathbf{b}, \mathbf{d}) \propto P(\theta) \prod_i \sum_{y_i} P(y_i|d_i) P(h_i|y_i, v_i, \theta), \tag{10.9}$$

and

$$P(y_i|\mathbf{b}, \mathbf{d}) \propto P(y_i|d_i) \sum_\theta P(\theta) \frac{P(h_i|y_i, v_i, \theta)}{P(h_i)} \prod_{j\neq i} P(y_j|d_j) P(h_j|y_j, v_j, \theta) \tag{10.10}$$

where the proportionality comes from excluded likelihood terms of the given **b** and **d**. If the object candidate is false, then its image height does not depend on the camera: $P(h_i|y_i = 0, v_i, \theta) = P(h_i)$. $P(h_i)$ is uniformly distributed in the range 0 to 1.

Now, the last step is to get $P(h_i|y_i = 1, v_i, \theta)$ from $P(Y_i)$. Let's assume that Y_i, the height of an object in a particular category, is normally distributed with parameters mean μ_i and standard deviation σ_i. Then, since $h_i = Y_i \frac{v_h - v_i}{Y_c}$, $P(h_i|y_i = 1, v_i, \theta)$ is normal with mean $\mu_i \frac{v_h - v_i}{Y_c}$ and standard deviation $\sigma_i \frac{v_h - v_i}{Y_c}$.

Now, we can put all of this together to improve object detection and recover the horizon and camera height, following the procedure below:

1. Provide priors for the horizon position v_h, the camera height Y_c, and the object physical size Y_i. These priors can be manually specified, or estimated from training data that contains ground truth object bounding boxes [104, 130].

2. Use an object detector to get $P(y_i|d_i)$ for each bounding box. If the detector outputs a score rather than a probability, the score can be converted to a probabilistic confidence by fitting a logistic function using a validation set [183].

3. Compute the likelihood of the objects and camera parameters using equations 10.9 and 10.10.

In this discussion, we have omitted just one annoying detail. The scores of strongly overlapping bounding boxes are not independent. A typical solution is to perform non-maxima suppression, removing all but the highest scoring candidate among boxes with substantial overlap. However, this solution is not ideal because it would remove candidates before considering the perspective evidence. Instead, Hoiem et al. [104] cluster the bounding boxes and encode candidates with high overlap as mutually exclusive alternatives, so that, e.g., $P(y_i|d_i)$ is replaced with $\sum_j P(y_i, k_i = j|d_{ij})$, where k_i is an indicator variable that specifies which bounding box within the cluster of candidates applies.

10.1.1.3 Results

To get a sense of the improvement that can result from modeling the size of objects, we show results from Hoiem et al. [104] in Figure 10.3. To get these results, Hoiem et al. used a data-driven method to estimate the horizon position, which they show led to some improvement. They also used the confidences of geometric context labels around the bounding boxes to refine the appearance scores. Most of the improvement is due to modeling perspective. In Figure 10.4, we show an automatically produced overhead view of the scene, as a proof-of-concept that modeling perspective will lead to better ways to reason about object relations.

10.1.2   APPEARANCE FEATURES

Typically, object detectors make an implicit weak perspective assumption: depth affects only size, not appearance. Such an assumption serves well for scene-scale images with distant cars and pedestrians, but it can lead to poor detection when objects are nearby and elongated. For example, when detecting

(a) Local Detection   (a) Full Model Detection   (b) Local Detection   (b) Full Model Detection

(e) Local Detection   (e) Full Model Detection   (f) Local Detection   (f) Full Model Detection

(g) Local Detection   (g) Full Model Detection   (h) Local Detection   (h) Full Model Detection

Figure 10.3: We show local object detections (left) of Dalal-Triggs (green=true car, cyan=false car, red=true ped, yellow=false ped) and the final detections (right) and horizon estimates (blue line) after considering surface geometry and camera viewpoint. The integrated approach yields large improvement (+7%/11% for peds/cars at 1 FP/image) over a very good local detector. Many of the remaining recorded "false positives" are due to objects that are heavily occluded (a,e) or very difficult to see (e) (i.e., missed by the ground truth labeler). In (h), a missed person on the right exemplifies the need for more robust assumptions (e.g., a person *usually* rests on the ground plane) or explanation-based reasoning (e.g., the person only looks so tall because he is standing on a step). Figure from [101].

tables or chairs in indoor scenes, small movements of the camera can cause large variations in appearance due to foreshortening and change in gradient orientations. If we know the perspective of the camera, we can be robust to such variations.

Hedau et al. [94] provide a good example of how to make better appearance models using perspective, illustrated in Figure 10.5. Their goal is to detect beds in rooms. They first detect three orthogonal vanishing points to determine the orientation of the room and the vanishing line of the ground plane. Then, they search for beds as axis-aligned 3D boxes, sliding the boxes along the floor, in a 3D analogue to sliding windows. As the box moves in the scene, the sides undergo perspective distortion to capture foreshortening. Also, rather than computing histograms of gradients in the

 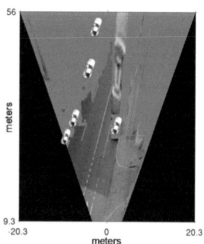

Figure 10.4: We project the estimated ground surface into an overhead view, using the estimated camera viewpoint, and plot scaled icons of objects (red dots for pedestrians) at their detected (using Dalal-Triggs [40]) ground positions. Car orientation is estimated by a robust line fit, assuming that cars mostly face down the same line. To plot in metric scale, we assume a typical focal length. Figure from [104].

image space, the image features are rectified along the room orientations. For example, a rectangle on the floor will have vertical and horizontal edges in the rectified features. In practice, the gradient features are rectified by weighting the magnitude based on the similarity of the gradient orientation and the angle to each vanishing point. Hedau et al. show that by combining the sliding cuboids and rectified features with more traditional 2D sliding window detectors, it is possible to improve accuracy and also to recover a better sense of the object position within the scene. Note that the Hedau et al. approach assumes that objects are aligned with the room. It would be straightforward to relax this by searching over a variety of 3D orientations, but a more relaxed model would also have more opportunity for false positives. Another interesting alternative would be to have a parts-based model in the spirit of Felzenszwalb et al. [66] or Bourdev and Malik [21], except that parts could be defined in the image plane or rectified to scene orientations.

10.1.3 INTERACTION BETWEEN OBJECTS AND SCENE VIA OBJECT SCALE AND POSE

So far the interaction between objects and scene has been modeled by focusing on the relationship between the camera and object size (scale) through a perspective transformation. A more generic interaction model can be formulated if one considers the explicit relationship among all of the object geometrical attributes (e.g., not just location and scale but also the pose of the object), the spatial properties of the scene (e.g., location and orientation of the planes where objects are grounded—

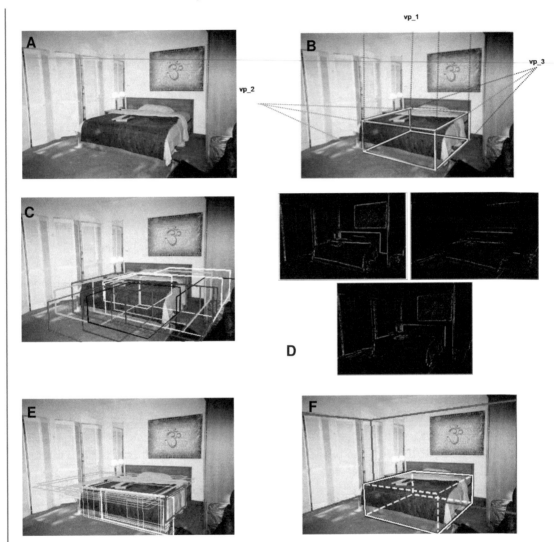

Figure 10.5: Overview of the "Thinking Inside the Box" algorithm [94]. From the input image (A), estimates of three orthogonal vanishing points are computed (B), providing the room's orientation. Object candidates (beds in this case; shown in C) are proposed by sliding a 3D cuboid of various sizes across the floor, in alignment with the room's walls. Object candidates are scored based on axis-aligned gradient features (D), computed by weighting gradient magnitudes by their alignment to each vanishing point. Each side of the object is scored separately; (E) shows high scoring samples for the front (yellow), top (green), and side (red) of the bed. Finally, using a probabilistic formulation, the object detections are refined by considering size and layout constraints with the room (borders of walls shown in red). The wireframe for the highest scoring bed detection is shown in yellow.

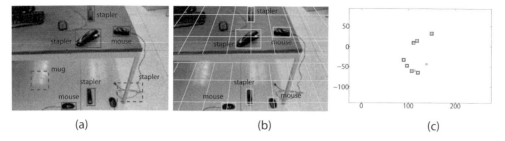

(a)   (b)   (c)

Figure 10.6: Results by the 3D layout estimator in Bao et al. [8]. (a) Example of baseline detections (marked by red bounding boxes) with false alarms (red dashed box). These are obtained using a baseline detector such as [222]. (b) Detections (marked by green bounding boxes) after the joint estimation process by the 3D layout estimator. The estimated supporting planes are superimposed in yellow. (c) The estimated 3D layout of the scene. The figure shows the side view of the 3D reconstructed scene (the camera is located at $(0, 0)$ pointing toward right). The estimated supporting plane is in green. Green dots are the objects detected and recovered in the 3D camera reference system by the 3D layout estimator; red squares are objects detected by the baseline detector.

the supporting planes), and the observer. These constraints enable the design of methods for 3D scene layout determination that require fewer assumptions on the camera model and the ability to use extended "ground plane models" (Chapter 3) wherein multiple supporting planes are allowed. Specifically, given a single uncalibrated image portraying the scene, the goal is to estimate the 3D scene layout and the camera parameters that best explains the observations. The observations are the object detection hypotheses and the estimation of their geometrical attributes (e.g., location, scale and pose). These observations can be returned by 3D object detectors (e.g., those based on the 2-1/2D layout or 3D layout models introduced in Part II). The 3D scene layout parameters are: 1) 3D location and pose, within the camera reference system, for each object; 2) orientation and distance from the camera of one or more planes that support the objects. The perspective camera parameters (Chapter 2) include the camera focal length and pose angles (e.g., tilt and roll). The scene and camera parameters are initially unknown and must be jointly inferred to select the most likely object hypotheses and compatible 3D layout.

Bao et al. [8] propose to formulate the joint inference process above as the following optimization problem:

$$\arg \max_{S, \{h_i\}} \ln p(o, E, S) = \arg \max_{S, \{h_i\}} p(S) + \sum_{i=1}^{N} [\ln p(h_i | a_i, S) + \ln p(e_i | h_i, a_i)] , \qquad (10.11)$$

where S collects unknown camera parameters and scene physical properties, $E = \{e_i\}$ describes the image evidence (i.e. image features used by the detectors to cast detection hypotheses) and $O = \{o_i\} = \{a_i, h_i\}$ collects the observations (list of detections hypotheses). h_i is a index associated to the semantic label of hypothesis i (car, person, bike, ..., none of above) and a_i describes its 3D geometrical attributes (position, scale and pose). Solving Eq. 10.11 is very challenging as the configuration space is large and exact inference is intractable. In order to simplify the solution of Eq. 10.11, Bao et al. [8] propose to incorporate geometrical constraints as penalty terms during the inference process and enforce priors on the distribution $p(S)$ (scene priors). Such geometrical constraints capture the relationship among objects' zenith poses ϕ, orientation and location of the supporting planes and the camera focal length f. It can shown that by measuring the zenith pose ϕ of at least three non-collinear objects, the focal length f and the normal of the supporting n can be estimated from a single image. Moreover, scene priors capture the (reasonable) assumption that supporting planes are mutually parallel and that objects are grounded up-right. This formulation has the advantage of enabling configurations where multiple supporting planes coexist and generic perspective camera models are used. Anecdotal results are reported in Figure 10.6. Note that unlike [101], the method proposed here leverages an additional critical piece of information—that is, measurements of the object zenith or azimuth pose angles. This concept is extended in Sun et al. [221] and further generalized in [7] where the case of multiple views of the same scene is studied.

10.2   SCENE LAYOUT

As Biederman notes [16], objects have a structured arrangement in well-formed scenes. Beds cannot stick through walls. People usually do not hang chairs from the ceiling. Chairs are placed on the floor, often near tables. Many approaches have been developed to model spatial relations in the image plane [92, 125, 126, 162, 186, 219, 243]. But objects are arranged according to *physical* laws and for convenience in *physical* interaction. We can best take advantage of their organization with some knowledge of the physical scene space. Hedau et al. [94] and Lee et al. [132] provide two examples of how to incorporate some simple layout principles, such as that objects cannot stick through walls and that beds are likely to have one side near a wall. Another good example comes from Cornelis et al. [36], where cars and people are detected by searching within the corridor formed by building walls. These detections, in turn, are used to refine estimates of the ground plane. Modeling priors and constraints of spatial layout to improve recognition or scene interpretation is an interesting topic of future work, particularly if combined with appearance models of contact, support, and occlusion.

10.3   OCCLUSION

Occlusion is one of the major challenges of object recognition. Most approaches to recognition ignore occlusion, hoping that classifiers will learn not to rely on features in parts of the object that are often occluded (e.g., the legs of people). With knowledge of the scene, we may be able to improve robustness to occlusion, but little work has been done in this area. Yang et al. [265] estimate depth order of segmented objects using shape priors but report no improvement in recognition. Hoiem et al. [104] reject object candidates that are not supported by occlusion boundaries, with mixed results. Endres and Hoiem [51] use estimated occlusion boundaries to effectively localize objects but do not consider categorization.

It's obvious that reasoning about occlusion should improve recognition, but we don't yet know how. Recovering occlusion boundaries and using them to improve recognition is an important research direction.

10.4   SUMMARY

In this chapter, we have discussed several ways to integrate object recognition and estimates of spatial layout. The size, position, viewpoint, and even appearance of an object depends on the scene layout and camera perspective. Thus, if we can recognize an object and its properties, we can predict aspects of the scene layout and vice versa. In Section 10.1.1, we discussed how to use perspective models to jointly improve confidences of 2D object detectors and to estimate the horizon position of the ground plane. In Section 10.1.2, we showed how estimated vanishing points can be used to guide appearance models for objects that are box-like and likely to be aligned with the walls, such as beds and other furniture. In Section 10.1.3, we discussed an approach to jointly infer object position, object pose, and the position and orientation of one or more supporting surfaces, integrating estimates of 3D object models and scene geometry. Beyond the works discussed in this chapter, we point the interested reader to works by Cornelis, Leibe, and others (e.g., [36, 136]). The problem of jointly reasoning about scene space and objects continues to be a highly active research topic.

CHAPTER 11

Cascades of Classifiers

In 1978, Marr proposed to organize the visual system as a sequential process, producing increasingly high-level descriptions of the scene: from a low-level primal sketch to a $2\frac{1}{2}$D sketch of surfaces to a full 3D model of the scene [156]. Unfortunately, with this model, a flaw in early processing can ruin the entire interpretation. Barrow and Tenenbaum [12] extended Marr's idea of geometric sketches to a general representation of the scene in terms of *intrinsic images*, each a registered map describing one characteristic of the scene. For example, the maps could indicate surface orientations, object boundaries, depth, or shading. But in contrast to Marr's feed-forward philosophy, Barrow and Tenenbaum proposed that the entire system should be organized around the recovery of these intrinsic images, so that they serve, not as a pre-process, but as an interface between the interdependent visual tasks. Their key idea is that the ambiguities of the scene can be resolved only when the many visual processes are working together.

In this chapter, we explore two algorithms that extend Barrow and Tenenbaum's intrinsic image framework. This framework has several advantages. First, it is synergistic, allowing all processes to benefit from their combined interaction. Second, the framework is modular, allowing a new process to be inserted without redesigning the entire system. Systems that define contextual relationships symbolically and perform inference over graphical models (e.g., [101, 126]) usually cannot easily accommodate new types of information. Third, by allowing one process to influence another through cues, rather than hard constraints (as in the original Barrow and Tenenbaum paper), the framework is robust and less sensitive to researcher-designed assumptions.

11.1 INTRINSIC IMAGES REVISITED

In this section, we describe an approach by Hoiem et al. [103] to integrate several scene analysis algorithms that describe surfaces [102] ("surface layout"), detect objects and infer viewpoint [101] ("objects in perspective"), and recover object occlusion boundaries and estimate depth [107] ("occlusion").[1] Each algorithm is treated as a component that takes as input the raw image and the intrinsic images from the other algorithms and outputs its own set of intrinsic images. The goal is to provide a more accurate and coherent scene interpretation by closing the feedback loop among them.

11.1.1 INTRINSIC IMAGE REPRESENTATION

The intrinsic images, displayed in Figure 11.2, serve as an interface between the various scene understanding processes. As proposed by Barrow and Tenenbaum [12], each intrinsic image is an

---
[1]Much of the text and figures for this section originally appeared in [103].

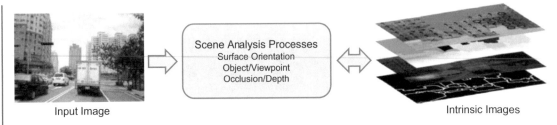

Figure 11.1: A simple framework for integrating disparate scene understanding processes using maps of scene characteristics as an interface. Figure from [103].

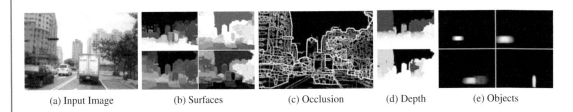

(a) Input Image   (b) Surfaces   (c) Occlusion   (d) Depth   (e) Objects

Figure 11.2: Examples of intrinsic images estimated from the image (a) in the first iteration. In (b), we show four of the surface confidence maps (brighter is higher confidence); clockwise, from upper-left: "support", "vertical planar", "vertical porous", "vertical solid". In (c), we show the confidence map for occlusion boundaries (bright indicates occlusion likely). In (d), we show upper and lower estimates of depth (log scale, brighter is closer). In (e), we show four of the object intrinsic images. Each is a confidence map indicating the likelihood of each pixel belonging to an individual object (cars or pedestrians in this case). Figure from [103].

image-registered map of one scene characteristic. These intrinsic images differ from those of Barrow and Tenenbaum in that they reflect the confidences of the estimates, either by representing the confidences directly, as with the surfaces, or by including several estimates, as with the depth.

Surfaces: The surface images consist of seven confidence maps for "support" (e.g., the ground), vertical planar facing "left", "center", or "right" (e.g., building walls), vertical non-planar "porous" (e.g., tree leaves), vertical non-planar "solid" (e.g., people), and "sky". Each image is a confidence map for one surface type indicating the likelihood that each pixel is of that type. The surface intrinsic images are computed using the algorithm described in Chapter 5, Section 5.1. The original algorithm, however, is modified by storing the multiple segmentations and augmenting the cue set with the contextual cues from the other processes.

Occlusions and Depth: Using the occlusion boundaries algorithm of Hoiem et al. [107], a boundary map that indicates likelihood of an occlusion boundary at each pixel and a set of depth maps are produced.

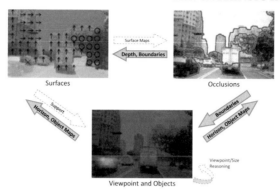

Figure 11.3: Contextual symbiosis. We show the final estimates for surfaces, occlusion boundaries, viewpoint, and objects and illustrate the interplay among them. The dotted arrows contain contextual relationships modeled by the previous work of Hoiem et al. [101, 107], while the solid arrows denote new cues proposed in this paper. For surfaces: green=support, red=vertical, blue=sky; arrows=planar orientation, X=solid, O=porous. Occlusion boundaries are denoted with blue/white lines. Objects are shown as overlaid individually colored confidence maps of object extent, with the blue line denoting the estimated horizon. Figure from [103].

Objects and Camera Viewpoint: Each hypothesized object is represented by a confidence map, indicating the likelihood that a pixel is part of the object times the likelihood that the object exists. The objects in perspective algorithm [101] (Chapter 10, Section 10.1.1) outputs a set of hypothesized objects, along with a probability distribution over potential bounding boxes. This distribution (using the Dalal-Triggs local detector [40]), along with an expected mask of the object class, is used to compute the likelihood that each pixel is part of an object instance. The expected mask of an object can be computed by averaging the masks of manually segmented objects in LabelMe [192]. The result is an "object map" for each object that is detected with confidence greater than some threshold. Objects at lower image positions are assumed to be closer to the viewer, so if two objects overlap, the confidences of the further object pixels are multiplied by one minus the confidences of the closer object pixels (loosely modeling object occlusion). The sum of the object maps over a pixel yields the likelihood that the pixel is generated by *any* detectable object. Intrinsic images are included for each hypothesis that passes a confidence threshold (5% in experiments). The camera viewpoint is represented with the camera height and the horizon position.

11.1.2 CONTEXTUAL INTERACTIONS

Here, we describe how the processes can interact using the intrinsic images as an interface between them. See Figure 11.3 for an illustration.

Surfaces and Objects: An object tends to correspond to a certain type of surface. For instance, the road is a supporting surface, and a pedestrian is a vertical, non-planar solid surface. In addition, many objects, such as cars and pedestrians tend to rest on the ground, so a visible supporting surface lends evidence to a hypothesized object. The values of the surface maps near a detection can strengthen or weaken the confidence in the detection (e.g., if the area below the detection is a supporting surface, confidence should increase). Likewise, because detected objects have a known geometry (e.g., "vertical non-planar solid"), the pixel labels of the objects can provide valuable cues for geometry estimation. Likewise, viewpoint estimates recovered from detected objects can improve surface estimates.

Surfaces and Occlusions: Occlusion boundaries often occur at the boundary between neighboring surface regions of different types. Further, the surface types are often a good indicator of the figure/ground label (e.g., the sky is always occluded). For this reason, cues based on surface estimates can greatly aid occlusion estimation. Additionally, the surface images, together with occlusion boundaries and camera viewpoint, are used to estimate depth, as in the original occlusion algorithm [107].

We can also use boundary and depth estimates to improve surface estimation. The boundary estimates help to provide better spatial support. In each segment produced by the multiple segmentation algorithm, the confidence of the most likely internal boundary provides a cue to whether a segment is likely to correspond to a single label. Also, the average depth and a measure of the slope in depth from left to right is computed each segment (for all three depth estimates). The average depth may help determine the label of the segment since appearance characteristics vary with distance (e.g., the texture in foliage is lost at a distance). The slope in depth may help determine whether a planar segment faces the left, center, or right of the viewer.

Objects and Occlusions: Object detections can aid occlusion reasoning by helping to determine whether neighboring regions are part of the same individual object. Thus, similarities and differences in the object pixel maps provide good occlusion features. Likewise, occlusion reasoning can help remove false object detections by showing them to be part of a larger structure. For example, pieces of the crosswalk in Figure 11.4 individually appear to be cars (and are consistent in viewpoint) but are discarded when found to be part of the larger cement structure.

11.1.3 TRAINING AND INFERENCE

Training and inference are performed in a simple iterative manner, cycling through the surface estimation, object detection and viewpoint recovery, and occlusion reasoning. We outline the training and inference algorithm in Figure 11.5. See [103] for implementation details.

11.1.4 EXPERIMENTS

Each algorithm is evaluated using the Geometric Context dataset. The first 50 images are used for training the surface segmentation and occlusion reasoning. The remaining 250 are used to test the surface, object, viewpoint, and occlusion estimators. The surface classifiers are trained and tested using five-fold cross-validation. We show examples of final results in Figure 11.6. In the object

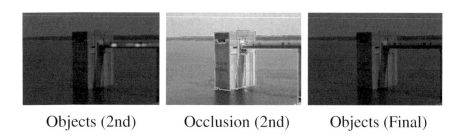

Objects (2nd)   Occlusion (2nd)   Objects (Final)

Figure 11.4: Example of the influence of occlusion on object estimates. Before considering occlusion information in the second iteration, pieces of the crosswalk are mistaken for cars (left). During the occlusion reasoning, however, it is determined that the cross-walk is a single structure, and the false detections are discarded. Figure from [103].

TRAINING for INTRINSIC IMAGES REVISITED

Initialize:
- Get multiple segmentations for each training image
- Estimate horizon
- Perform local object detection

For iteration $t = 1..N_t$:
1. Train and apply surface estimation
   (a) Compute features for each segment (using results of (2),(3) from iterations $1 \ldots t - 1$)
   (b) Train surface classifiers and compute surface confidences with cross-validation
2. Apply object/viewpoint/surface inference (using result of (1) from iteration t and (3) from $t - 1$)
3. Train and apply occlusion reasoning algorithm (using results of (1), (2) from iteration t)
   (a) Train on hold-out set
   (b) Perform occlusion reasoning on cross-validation images

---

Figure 11.5: Iterative training algorithm for combining surface, occlusion, viewpoint, and object information. Training and testing is performed on the Geometric Context dataset. The holdout set of 50 images used to train the surface segmentation algorithm is used to train the occlusion reasoning, and the remaining 250 images are used for testing (using five-fold cross-validation for the surface estimation). N_t =3 in Hoiem et al.'s experiments [103].

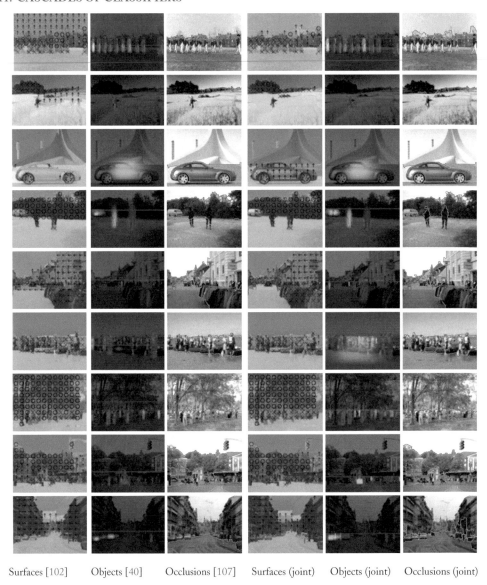

Surfaces [102]   Objects [40]   Occlusions [107]   Surfaces (joint)   Objects (joint)   Occlusions (joint)

Figure 11.6: In each row, we show the results of the three original algorithms and the results when they are integrated with Hoiem et al.'s framework [103]. Each process achieves higher performance when they work together. Rows 1 and 2: the occlusion estimates help improve mistakes in the surface labels. Row 3: the detected car allows the surface labels and occlusion estimate to improve. Row 4: the soldiers need better camouflage as the algorithm is able to identify and segment them away from the background. Row 5: false detections on the coat rack are eliminated due to the occlusion information. Row 6: The contextual inference results in more pedestrians being detected (with a false positive car). Rows 7-9: A reasonably good job is done in complicated scenes. Figure from [103].

results here and elsewhere, only objects that pass a preset confidence threshold are shown (threshold corresponds to 0.5 false positives per image).

In this subsection, we discuss the improvement in each type of estimation. Our analysis includes qualitative assessment, inspection of the decision tree classifiers learned in the surface and occlusion estimation, and quantitative performance comparison. In the decision tree learning, early and frequent selection of a cue indicates that the cue is valuable but does not necessarily imply great value beyond the other cues, as there may be redundant information.

Surfaces: The decision tree learning indicates that the boundary likelihood from occlusion reasoning is the most powerful cue for the segmentation classifier, as it is the first and most frequently selected. For the "solid" classifier, the pedestrian confidence value from the object detection is the first feature selected. The learning algorithm determines that regions with high pedestrian confidence are very likely to be solid, but regions without pedestrian confidence are not much less likely (i.e., all pedestrians are solid, but not all solids are pedestrians). The measure of depth slope from the occlusion reasoning is used frequently by the planar "left" and "right" classifiers.

Also, the position cues relative to the estimated horizon position are used overall twice as frequently as the absolute position cues. In a separate experiment, in which surface classification was trained and tested with manually assigned (ground truth) horizon estimates, the main (support, vertical, sky) classification and subclassification of vertical (left, center, right, porous, solid) each improve by 2% (from 87% to 89% and 61% to 63%, respectively). Thus, knowledge of the horizon is an important cue, but the potential improvement in surface estimation by improving the horizon estimate is limited.

Inclusion of object and occlusion information yields a modest quantitative improvement in surface classification. Using the surface layout algorithm [99] by itself, the main classification accuracy is 86.8% and subclassification accuracy is 60.9%. The accuracy improves by roughly 1% to 87.6% and 61.8% once the contextual information from the objects, occlusions, and depth is considered. For the main classification ("support" vs. "vertical" vs. "sky"), this difference is statistically significant ($p < 0.05$), but for the subclassification (subclasses within "vertical") it is not ($p > 0.05$). With closer inspection on an image-by-image basis, of the 18% of images that change in main class pixel error by more than 5%, 74% improve. Thus, while large changes are rare, changes made after considering the new object and occlusion information are much more likely to improve results than not.

Objects: The Dalal-Triggs [40] object detector is used to supply object hypotheses and confidences based on local image information. When viewpoint and surfaces are considered (result of the first iteration, equivalent to the objects in perspective algorithm [101]), pedestrian detection improves considerably. When occlusion information is considered (later iterations) car detection improves but pedestrian detection drops slightly, likely due to the difficulty of maintaining occlusion boundaries for distant pedestrians (see Figure 11.7). Along with many false positives, 11% of true pedestrian and 8% of true car detections are discarded by the occlusion-based filtering. Overall, the car detection improves by 10% and the pedestrian detection by 7% when considering surface, viewpoint, and occlusions.

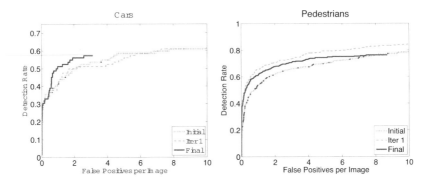

Figure 11.7: ROC curves for initial Dalal-Triggs classification, after one iteration (equivalent to the "Putting Objects in Perspective" algorithm [101]), and the final result. Figure from [103].

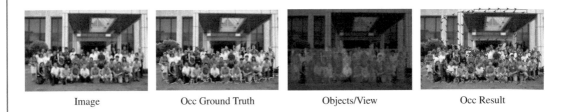

Figure 11.8: By reasoning together about objects and occlusions, the computer is sometimes able to find the occlusion boundaries of tightly crowded individuals. In this case, the occlusion estimates are more precise than the ground truth, which leads to an artificially low quantitative measure of improvement. Figure from [103].

Viewpoint: Viewpoint estimation is evaluated based on the horizon position, since it is difficult to obtain ground truth for camera height. Using the mean horizon position as a constant estimate yields an error of 12.8% (percentage of image height difference between the manually labeled horizon and the estimated horizon). This error drops to 10.4% when using a data-driven horizon estimate with the LabelMe training set. During the first iteration, which is equivalent to the objects in perspective algorithm [101], this drops further to 8.5%. Further iterations do not produce a statistically significant change ($p > 0.05$).

Occlusion: A subjective comparison reveals that individual cars and people are correctly delineated much more frequently when object information is considered. Figure 11.6 contains many such examples. However, Hoiem et al. are not able to measure an overall quantitative improvement, due to ground truth labeling of a crowd of people or row of cars as a single object, as shown in Figure 11.8.

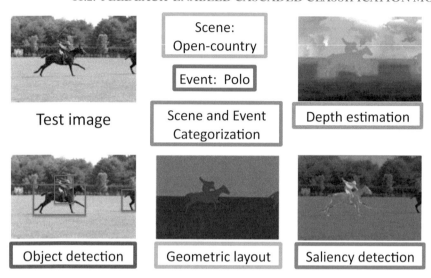

Figure 11.9: Li et al. [141] aim to learn to recognize a variety of scene properties while taking advantage of their correlations. Detectable properties include scene category, event category, depth maps (Black = close, white = far), object bounding boxes (red = horse, blue = person), surface layout maps (green = vertical-porous, red = horizontal, blue = vertical) and salience (cyan = salient).

11.2   FEEDBACK-ENABLED CASCADED CLASSIFICATION MODELS

Here, we describe an elegant framework proposed by Li et al. [141] for creating a feedback loop between scene understanding processes, which builds on the earlier work of Heitz et al. [96]. As in the last section, each algorithm produces features that feed into the others. However, in this approach, called Feedback-Enabled Cascaded Classification Models or FE-CCM, the algorithms are trained jointly, so that the final outputs are as good as possible. The FE-CCM approach does not depend on the details of the individual algorithms, their features, or their datasets, making it easy to integrate multiple tasks (Figure 11.9). Each algorithm can be trained and tested on a separate dataset, while still benefitting from the other trained algorithms. In experiments, the authors synergistically combine the tasks of event categorization, depth estimation, scene categorization, salience detection, geometric labeling, and object detection. Impressively, they demonstrate improved performance (substantially, in most cases) in each task.

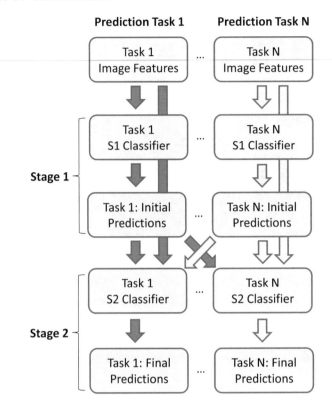

Figure 11.10: Illustration of the FE-CCM algorithm. For each scene prediction task (e.g., categorize scene, estimate depth, label objects), two classifiers are trained. The first classifier uses image features to try to predict some target value (not necessarily ground truth). The second uses both image features and the output from all first stage classifiers to predict the ground truth labels. In training, the parameters of each classifier and the target predictions for the first stage are learned together.

11.2.1 ALGORITHM

The goal is to jointly estimate several scene properties, including scene categories, surface labels, depth, and object locations. See Figure 11.10 for an illustration of the algorithm. The CCM-FE approach is to break down the estimation into a two-stage cascade:

- **Stage 1:** For each element, a classifier makes predictions based on features computed from the image. There are no constraints on the form of the classifier or the features (though a linear classifier is easier to train).

- **Stage 2:** For each element, a classifier makes predictions using both the image features and features computed from the stage 1 classifiers.

There are three phases to training:

- **Initialization:** The "target" confidences for each label of the stage 1 classifiers is set according to the ground truth. For example, if the scene is "forest", then the scene categorizer will ideally predict "forest" with likelihood of 1.

- **Feed-forward Training:** The stage 1 classifiers are trained to try to predict their target confidences from image features. Stage 2 classifiers are trained to predict ground truth labels from image features and features computed from the output of the stage 1 classifiers.

- **Feedback Training:** Keeping the classifier parameters fixed, new targets are set for the stage 1 classifiers that are close to the current stage 1 predictions but lead to more accurate predictions from the stage 2 classifiers (see [141] for technical details).

The training iterates between the feed-forward and feedback phases. To make predictions for a new image, the stage 1 classifiers are applied, followed by the stage 2 classifiers.

11.2.2 EXPERIMENTS

Li et al. [141] report experiments using the following scene elements: scene categories, depth estimates, event categories, salience, object locations, and geometric classes. The categories are represented as vectors of log-odds scores (the log likelihood of each class minus the log likelihood of not that class) or classifier scores. The depth estimates are continuous estimates for each pixel. Salience, a measure of an image region's likelihood to attract attention, is predicted as a thresholded score. For objects, a set of bounding boxes are scored and thresholded to determine whether they are true detections. For the geometric classes, only the main classes ("support", "vertical", "sky") were considered. Experiments were performed by training on the original datasets that were proposed with the individual tasks. Typically, Li et al. use the original or state-of-the-art algorithm for the first stage of each task and a logistic regression classifier for the second stage (which makes the training easier). The outputs from the first stage classifiers were used directly as features for the second stage classifiers (there may have been some modification to align features with the object windows, but it is not reported).

Examples of results are shown in Figure 11.11. Li et al. show that performance increases for each of the tasks from the first stage to the second stage: event category accuracy (71.8% to 74.3%); depth RMSE in meters (16.7 to 15.5); scene category accuracy (83.8% to 85.9%); salience detection accuracy (85.2% to 86.2%); geometric labeling accuracy (86.2% to 88.6%); and object detection average precision (0.444 to 0.454). One particularly noteworthy result is that the algorithm achieves higher accuracy than the surface layout and Make3D algorithms (Chapter 5) in estimating geometric labels and depth, respectively, even though this algorithm did not use multiple segmentations or MRFs. The long-range spatial support and interactions provided by the multiple segmentations and MRFs seems to be sufficiently provided by the outputs of the stage 1 classifiers.

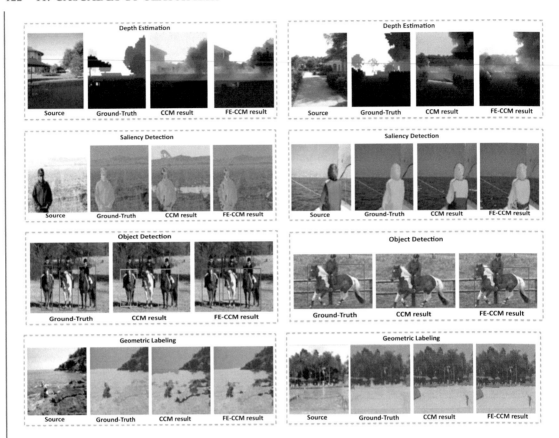

Figure 11.11: Results from the CCM method of Heitz et al. [96] and the FE-CCM method of Li et al. [141]. First row: depth maps in the first row are at the same scale (black is near, white is far). Second row: salient regions are indicated in cyan. Third row: bounding boxes are drawn around detected horses. Fourth row: green = "support", blue = "sky", and red = "vertical". Figure from an extended version of [141].

11.3   SUMMARY

We have discussed two approaches to represent scene properties with pixel maps, as first proposed by Barrow and Tenenbaum [12]. Hoiem et al. [103] jointly label surface geometry, occlusion boundaries, objects, and camera viewpoint, iteratively passing features computed from each properties map to the classifiers that update estimates of the other properties. Li et al. [141] provide an elegant framework that reduces the inter-task feature design and facilitates use of a variety of datasets with incomplete annotations.

CHAPTER 12

Conclusion and Future Directions

Our aim in this book is to help the reader to build a foundational understanding of 3D scene interpretation and object recognition in images: the mathematical underpinnings, its early roots, and its revival and incorporation of statistical machine learning. We hope that the reader will be inspired to continue on to make progress in this area. Currently, research on 3D scene and object models is in a brainstorming phase, with many proposed models but no consensus on a "correct" approach. As such, open problems abound. We conclude by discussing a few important directions:

- *Integrating single-view and multi-view approaches:* Efforts to model geometry from two or more images [91] rely on matching points across the images and tend to produce unstructured 3D point clouds. Single-view methods tend to make strong assumptions about the scene geometry and use color, texture, and perspective cues with learned appearance models. We would like to recover both the detailed geometry and abstractions into major surfaces and objects from a limited number of images, which will likely require building on efforts of both the single-view and multi-view communities. See [7] for one recent effort in this direction.

- *Taking advantage of structure without relying on it:* We currently have different models for cluttered indoor scenes, hallways, streets, and natural outdoor environments. We would like a single system to be able to represent a wide variety of scenes, making use of structured priors where appropriate but also capable of handling more general geometries. Such a system may switch between several different representations (sometimes called schemas) or perhaps use a flexible grammar.

- *Interpreting the physical components of scenes:* Despite some recent work on detecting shadows [82, 129, 270] in consumer photographs, estimating illumination [128], and labeling surface geometries [102], these problems are still far from solved. We would like to be able to represent large-scale shapes (3D rectangles, curved surfaces, etc.) and small-scale geometries (divots, bumps, bends, corners, etc.) and albedo, requiring integrated reasoning about surface orientation and boundaries, illumination, and materials.

- *Integrating Different Sensing Modalities* As different types of sensors are becoming widely accessible to consumers at large (e.g., sensors such as those by PrimeSense (www.primsense.com) provide RGB-valued images along with a registered depth map of the scene), we would like to coherently combine various types of sensing modalities (e.g., imagery, depth maps, etc.)

into a coherent inference engine that provides robust estimates of the scene layout. The works by [25, 217] are recent examples of such frameworks.

Bibliography

[1] E. H. Adelson. On seeing stuff: The perception of materials by humans and machines. In *Proceedings of the SPIE*, 2001. DOI: 10.1117/12.429489 Cited on page(s) 17

[2] S. Agarwal, N. Snavely, I. Simon, S. M. Seitz, and R. Szeliski. In *ICCV*, 2009. Cited on page(s) xviii, 20

[3] P. Arbelaez, M. Maire, C. Fowlkes, and J. Malik. From contours to regions: An empirical evaluation. In *CVPR*, 2009. DOI: 10.1109/CVPR.2009.5206707 Cited on page(s) 30

[4] M. Arie-Nachimson and R. Basri. Constructing implicit 3d shape models for pose estimation. In *ICCV*, 2009. Cited on page(s) 70, 75, 77, 79, 80, 81

[5] J. S. Bakin, K. Nakayama, and C. D. Gilbert. Visual responses in monkey areas v1 and v2 to three-dimensional surface configurations. *The Journal of Neuroscience*, 20(21):8188–8198, Nov. 2000. Cited on page(s) 18

[6] D. H. Ballard. Generalizing the hough transform to detect arbitrary shapes. *Pattern Recognition*, 13(2):111–122, 1981. DOI: 10.1016/0031-3203(81)90009-1 Cited on page(s) 62, 70

[7] S. Y. Bao and S. Savarese. Semantic structure from motion. In *CVPR*, 2011. Cited on page(s) 108, 123

[8] S. Y. Bao, M. Sun, and S. Savarese. Toward coherent object detection and scene layout understanding. In *CVPR*, 2010. DOI: 10.1109/CVPR.2010.5540229 Cited on page(s) 20, 22, 107, 108

[9] A. Barbu and S. Zhu. Generalizing Swendsen-Wang to sampling arbitrary posterior probabilities. *IEEE Trans. on Pattern Analysis and Machine Intelligence*, 27(8):1239–1253, 2005. DOI: 10.1109/TPAMI.2005.161 Cited on page(s) 8

[10] O. Barinova, V. Konushin, A. Yakubenko, K. Lee, H. Lim, and A. Konushin. Fast automatic single-view 3-d reconstruction of urban scenes. In *ECCV*, 2008. DOI: 10.1007/978-3-540-88688-4_8 Cited on page(s) 8, 23

[11] O. Barinova, V. Lempitsky, E. Tretiak, and P. Kohli. Geometric image parsing in man-made environments. In *ECCV*, 2010. DOI: 10.1007/978-3-642-15552-9_5 Cited on page(s) 15

[12] H. Barrow and J. Tenenbaum. Recovering intrinsic scene characteristics from images. In *Comp. Vision Systems*, 1978. Cited on page(s) 4, 18, 19, 21, 111, 122

[13] H. Barrow and J. Tenenbaum. Interpreting line drawings as three-dimensional surfaces. *Artificial Intelligence*, 17:75–116, 1981. DOI: 10.1016/0004-3702(81)90021-7 Cited on page(s) 6

[14] E. Bart, E. Byvatov, and S. Ullman. View-invariant recognition using corresponding object fragments. In *ECCV*, pages 152–165, 2004. DOI: 10.1007/978-3-540-24671-8_12 Cited on page(s) 71, 79, 81

[15] T. L. Berg, A. Sorokin, G. Wang, D. A. Forsyth, D. Hoiem, A. Farhadi, and I. Endres. It's all about the data. *Proceedings of the IEEE, Special Issue on Internet Vision*, 98(8):1434–1453, August 2010. DOI: 10.1109/JPROC.2009.2032355 Cited on page(s) 32

[16] I. Biederman. On the semantics of a glance at a scene. In M. Kubovy and J. R. Pomerantz, editors, *Perceptual Organization*, chapter 8. Lawrence Erlbaum, 1981. Cited on page(s) xvii, 5, 108

[17] I. Biederman. Human image understanding: Recent research and theory. *Computer Vision, Graphics and Image Understanding*, 32:29–73, 1985. Cited on page(s) 57

[18] I. Bierderman. Recognition-by-components: A theory of human image understanding. *Psychological Review*, 94(2):115–147, 1987. DOI: 10.1037/0033-295X.94.2.115 Cited on page(s) 57

[19] T. Binford. Visual perception by computer. *IEEE conference on Systems and Control*, 1971. Cited on page(s) 58

[20] D. M. Blei. Variational methods for the dirichlet process. In *ICML*, 2004. DOI: 10.1145/1015330.1015439 Cited on page(s) 80, 90

[21] L. Bourdev and J. Malik. Poselets: Body part detectors trained using 3d human pose annotations. In *ICCV*, 2009. DOI: 10.1109/ICCV.2009.5459303 Cited on page(s) 105

[22] K. W. Bowyer and C. R. Dyer. Aspect graphs: An introduction and survey of recent results. In *International Journal of Imaging Systems and Technology*, volume 2, pages 315–328 vol.2, 1990. DOI: 10.1002/ima.1850020407 Cited on page(s) 62

[23] L. Breiman. Random forests. *Machine Learning*, 45(1):5–32, 2001. DOI: 10.1023/A:1010933404324 Cited on page(s) 90

[24] R. Brooks, R. Greiner, and T. Binford. Model-based three-dimensional interpretation of two-dimensional images. In *IJCAI*, 1979. DOI: 10.1109/TPAMI.1983.4767366 Cited on page(s) 7

[25] G. J. Brostow, J. Shotton, J. Fauqueur, and R. Cipolla. Segmentation and recognition using structure from motion point clouds. In *ECCV*, pages 44–57, 2008. DOI: 10.1007/978-3-540-88682-2_5 Cited on page(s) 124

[26] M. Brown and D. G. Lowe. Unsupervised 3d object recognition and reconstruction in unordered datasets. In *International Conference on 3-D Digital Imaging and Modeling*, pages 56–63, Washington, DC, USA, 2005. IEEE Computer Society. DOI: 10.1109/3DIM.2005.81 Cited on page(s) 68, 83

[27] P. Cavanagh. The artist as a neuroscientist. *Nature*, 434:301–307, 2005. DOI: 10.1038/434301a Cited on page(s) 19

[28] S. Chen and L. Williams. View interpolation for image synthesis. *Computer Graphics*, 27:279–288, 1993. DOI: 10.1145/166117.166153 Cited on page(s) 85

[29] H. Chiu, L. Kaelbling, and T. Lozano-Perez. Virtual training for multi-view object class recognition. In *CVPR*, 2007. Cited on page(s) xix, 73, 75, 79

[30] H. Chiu, H. Liu, L. Kaelbling, and T. Lozano-Perez. Class-specific grasping of 3d objects from a single 2d image. In *IROS*, 2010. DOI: 10.1109/CVPR.2007.383044 Cited on page(s) xix, 73, 75, 95

[31] K. Chua and M. Chun. Implicit scene learning is viewpoint dependent. *Perception & Psychophysics*, 65(1):72–80, 2003. DOI: 10.3758/BF03194784 Cited on page(s) 5

[32] O. Chum, J. Philbin, J. Sivic, M. Isard, and A. Zisserman. Total recall: Automatic query expansion with a generative feature model for object retrieval. pages 1–8. IEEE, 2007. DOI: 10.1109/ICCV.2007.4408891 Cited on page(s) 66

[33] A. Collet Romea, D. Berenson, S. Srinivasa, and D. Ferguson. Object recognition and full pose registration from a single image for robotic manipulation. In *ICRA*, 2009. DOI: 10.1109/ROBOT.2009.5152739 Cited on page(s) 68, 81

[34] M. Collins, R. Schapire, and Y. Singer. Logistic regression, Adaboost and Bregman distances. *Machine Learning*, 48(1–3), 2002. DOI: 10.1023/A:1013912006537 Cited on page(s) 41

[35] D. Comaniciu and P. Meer. Mean shift: A robust approach toward feature space analysis. *IEEE Trans. on Pattern Analysis and Machine Intelligence*, 24:603–619, May 2002. DOI: 10.1109/34.1000236 Cited on page(s) 29, 68

[36] N. Cornelis, B. Leibe, K. Cornelis, and L. Gool. 3d urban scene modeling integrating recognition and reconstruction. *International Journal of Computer Vision*, 78:121–141, July 2008. DOI: 10.1007/s11263-007-0081-9 Cited on page(s) 108, 109

[37] A. Criminisi, I. Reid, and A. Zisserman. Single view metrology. *International Journal of Computer Vision*, 40(2), 2000. DOI: 10.1023/A:1026598000963 Cited on page(s) xvii, 10, 13

[38] J. E. Cutting and P. M. Vishton. Perceiving layout and knowing distances: The integration, relative potency, and contextual use of different information about depth. In W. Epstein and S. Rogers, editors, *Perception of Space and Motion*, pages 69–117. Academic Press, San Diego, 1995. Cited on page(s) 4

[39] C. M. Cyr and B. B. Kimia. A similarity-based aspect-graph approach to 3d object recognition. *International Journal of Computer Vision*, 57(1):5–22, 2004. DOI: 10.1023/B:VISI.0000013088.59081.4c Cited on page(s) 62

[40] N. Dalal and B. Triggs. Histograms of oriented gradients for human detection. In *CVPR*, 2005. DOI: 10.1109/CVPR.2005.177 Cited on page(s) 35, 102, 105, 113, 116, 117

[41] C. Dance, J. Willamowski, L. Fan, C. Bray, and G. Csurka. Visual categorization with bags of keypoints. In *ECCV International Workshop on Statistical Learning in Computer Vision.*, Prague, 2004. DOI: 10.1234/12345678 Cited on page(s) 83

[42] E. Delage, H. Lee, and A. Y. Ng. A dynamic Bayesian network model for autonomous 3D reconstruction from a single indoor image. In *CVPR*, 2006. DOI: 10.1109/CVPR.2006.23 Cited on page(s) 8, 42

[43] J. Deng, W. Dong, R. Socher, L.-J. Li, K. Li, and L. Fei-Fei. Imagenet: A large-scale hierarchical image database. In *CVPR*, 2009. DOI: 10.1109/CVPR.2009.5206848 Cited on page(s) 77

[44] S. J. Dickinson, A. P. Pentland, and A. Rosenfeld. 3-d shape recovery using distributed aspect matching. *IEEE Trans. on Pattern Analysis and Machine Intelligence*, 14(2):174–198, 1992. DOI: 10.1109/34.121788 Cited on page(s) 62

[45] R. Duda and P. Hart. Use of the hough transformation to detect lines and curves in pictures. *Communications of the ACM*, 15(1):11–15, 1972. DOI: 10.1145/361237.361242 Cited on page(s) 42

[46] R. Duda, P. Hart, and D. Stork. *Pattern Classification*. Wiley-Interscience Publication, 2000. Cited on page(s) 32, 34

[47] M. e. Everingham. The 2005 pascal visual object class challenge. in selected proceedings of the 1st pascal challenges workshop, to appear. Cited on page(s) 70, 71, 72

[48] A. Efros and A. Torralba. Unbiased look at dataset bias. In *CVPR*, 2011. Cited on page(s) 33

[49] D. Eggert and K. Bowyer. Computing the perspective projection aspect graph of solids of revolution. *IEEE Trans. Pattern Anal. Mach. Intell.*, 15(2):109–128, 1993. DOI: 10.1109/34.192483 Cited on page(s) 62

[50] D. Eggert, K. Bowyer, C. Dyer, H. Christensen, and D. Goldgof. The scale space aspect graph. *IEEE Transactions on Pattern Analysis and Machine Intelligence*, 15(11):1114–1130, 1993. DOI: 10.1109/34.244674 Cited on page(s) 62

[51] I. Endres and D. Hoiem. Category independent object proposals. In *ECCV*, 2010. Cited on page(s) 108

[52] A. Ess, B. Leibe, and L. Van Gool. Depth and appearance for mobile scene analysis. In *ICCV*, 2007. DOI: 10.1109/ICCV.2007.4409092 Cited on page(s) 99

[53] M. Everingham, L. Van Gool, C. K. I. Williams, J. Winn, and A. Zisserman. The PAS-CAL Visual Object Classes Challenge 2007 (VOC2007) Results. http://www.pascal-network.org/challenges/VOC/voc2007/workshop/index.html. DOI: 10.1007/s11263-009-0275-4 Cited on page(s) 33

[54] M. Everingham, L. Van Gool, C. K. I. Williams, J. Winn, and A. Zisserman. The PAS-CAL Visual Object Classes Challenge 2008 (VOC2008) Results. http://www.pascal-network.org/challenges/VOC/voc2008/workshop/index.html. Cited on page(s) 77

[55] M. Everingham, L. Van Gool, C. K. I. Williams, J. Winn, and A. Zisserman. The PAS-CAL Visual Object Classes Challenge 2010 (VOC2010) Results. http://www.pascal-network.org/challenges/VOC/voc2010/workshop/index.html. Cited on page(s) 77

[56] M. Everingham, A. Zisserman, C. K. I. Williams, and L. Van Gool. The PASCAL Visual Object Classes Challenge 2006 (VOC2006) Results. http://www.pascal-network.org/challenges/VOC/voc2006/results.pdf. Cited on page(s) 77, 91, 92, 93

[57] A. Farhadi, M. Hejrati, A. Sadeghi, P. Young, C. Rashtchian, J. Hockenmaier, and D. Forsyth. Every picture tells a story: Generating sentences for images. In *ECCV*, 2010. Cited on page(s) 19

[58] A. Farhadi, M. Kamali, I. Endres, and D. Forsyth. A latent model of discriminative aspect. In *ICCV*, 2009. Cited on page(s) 72

[59] A. Farhadi and A. Sadeghi. Recognition using visual phrases. In *CVPR*, 2011. Cited on page(s) 19

[60] O. Faugeras and M. Hebert. The representation, recognition and locating of 3d objects. In *International Journal of Robotic Research*, 1986. DOI: 10.1177/027836498600500302 Cited on page(s) 62

[61] O. Faugeras, M. Hebert, E. Pauchon, and J. Ponce. Object representation, identification, and positioning from range data. In *1st International Symposium on Robotics Research*, pages 425–446. Cited on page(s) 62

[62] L. Fei-Fei, R. Fergus, and A. Torralba. Recognizing and learning object categories. *CVPR 2007 Short Course*. Cited on page(s) 68

[63] P. Felzenszwalb, R. B. Girshick, D. A. McAllester, and D. Ramanan. Object detection with discriminatively trained part-based models. *IEEE Trans. on Pattern Analysis and Machine Intelligence*, 32(9):1627–1645, 2010.
DOI: 10.1109/TPAMI.2009.167 Cited on page(s) 68, 69

[64] P. Felzenszwalb and D. Huttenlocher. Pictorial structures for object recognition. In *IEEE. Computer Vision and Pattern Recognition.*, pages 2066–2073, 2000.
DOI: 10.1023/B:VISI.0000042934.15159.49 Cited on page(s) 68, 69, 84

[65] P. Felzenszwalb and D. Huttenlocher. Efficient graph-based image segmentation. *International Journal of Computer Vision*, 59(2), 2004. DOI: 10.1023/B:VISI.0000022288.19776.77 Cited on page(s) 29, 41, 47

[66] P. Felzenszwalb, D. McAllester, and D. Ramanan. A discriminatively trained, multiscale, deformable part model. In *CVPR*, 2008. DOI: 10.1109/CVPR.2008.4587597 Cited on page(s) 35, 80, 105

[67] R. Fergus, P. Perona, and A. Zisserman. Object class recognition by unsupervised scale-invariant learning. In *CVPR*, volume 2, pages 264–271, 2003.
DOI: 10.1109/CVPR.2003.1211479 Cited on page(s) 68, 84

[68] V. Ferrari, T. Tuytelaars, and L. Gool. Simultaneous object recognition and segmentation from single or multiple model views. *International Journal of Computer Vision*, 67(2):159–188, 2006. DOI: 10.1007/s11263-005-3964-7 Cited on page(s) 65, 66, 70, 71, 83

[69] V. Ferrari, T. Tuytelaars, and L. V. Gool. Integrating multiple model views for object recognition. In *CVPR*, 2004. DOI: 10.1109/CVPR.2004.1315151 Cited on page(s) 66, 70

[70] M. Fischler and R. Bolles. Random sample consensus: A paradigm for model fitting with applications to image analysis and automated cartography. In *Comm. of the ACM.*, volume 24, pages 381–395, 1981. DOI: 10.1145/358669.358692 Cited on page(s) 62, 68, 81

[71] D. Forsyth, J. Mundy, A. Zisserman, C. Coelho, A. Heller, and C. Rothwell. Invariant descriptors for 3d object recognition and pose. *IEEE Transactions on Pattern Analysis and Machine Intelligence*, 13, 1991. DOI: 10.1109/34.99233 Cited on page(s) 62

[72] D. A. Forsyth and J. Ponce. *Computer Vision: A Modern Approach*. Prentice Hall, 2002. Cited on page(s) xi, 62

[73] W. T. Freeman and E. H. Adelson. The design and use of steerable filters. *IEEE Trans. on Pattern Analysis and Machine Intelligence*, 13:891–906, September 1991. DOI: 10.1109/34.93808 Cited on page(s) 35

[74] W. T. Freeman and J. B. Tenenbaum. Learning bilinear models for two-factor problems in vision. In *CVPR*, 1997. DOI: 10.1109/CVPR.1997.609380 Cited on page(s) 72

[75] J. Friedman, T. Hastie, and R. Tibshirani. Additive logistic regression: a statistical view of boosting. *Annals of Statistics*, 28(2), 2000. DOI: 10.1214/aos/1016218223 Cited on page(s) 32

[76] J. Gibson. *The Perception of the Visual World*. Houghton Mifflin, Boston, USA, 1950. Cited on page(s) 3

[77] I. Gordon and D. G. Lowe. What and where: 3d object recognition with accurate pose. In *Toward Category-Level Object Recognition*, pages 67–82, 2006. DOI: 10.1007/11957959_4 Cited on page(s) 68, 81

[78] C. V. Gottesman. Viewpoint changes affect priming of spatial layout. *Journal of Vision*, 3(9), 2003. DOI: 10.1167/3.9.641 Cited on page(s) 5

[79] G. Griffin, A. Holub, and P. Perona. Caltech-256 object category dataset. Technical Report 7694, California Institute of Technology, 2007. Cited on page(s) 77

[80] C. Gu and X. Ren. Discriminative mixture-of-templates for viewpoint classification. In *ECCV*, pages 408–421, 2010. DOI: 10.1007/978-3-642-15555-0_30 Cited on page(s) 69, 78, 79, 80

[81] C. Guo, S. Zhu, and Y. N. Wu. Towards a mathematical theory of primal sketch and sketchability. In *ICCV*, 2003. Cited on page(s) 8

[82] R. Guo, Q. Dai, and D. Hoiem. Single-image shadow detection and removal using paired regions. In *CVPR*, 2011. Cited on page(s) 123

[83] A. Gupta and L. S. Davis. Objects in action: An approach for combining action understanding and object perception. In *CVPR*, 2007. DOI: 10.1109/CVPR.2007.383331 Cited on page(s) 19, 99

[84] A. Gupta, A. A. Efros, and M. Hebert. Blocks world revisited: Image understanding using qualitative geometry and mechanics. In *ECCV*, 2010. DOI: 10.1007/978-3-642-15561-1 Cited on page(s) xviii, 8, 20, 23, 25, 30

[85] A. Gupta, S. Satkin, A. A. Efros, and M. Hebert. From 3d scene geometry to human workspace. In *CVPR*, 2011. Cited on page(s) 25

[86] A. Guzman. Computer recognition of three-dimensional objects in a visual scene. Technical Report MAC-TR-59, MIT, 1968. Cited on page(s) 7

[87] F. Han and S. Zhu. Bayesian reconstruction of 3D shapes and scenes from a single image. In *Int. Work. on Higher-Level Know. in 3D Modeling and Motion Anal.*, 2003. Cited on page(s) 8

[88] F. Han and S. Zhu. Bottom-up/top-down image parsing by attribute graph grammar. In *ICCV*, 2005. DOI: 10.1109/ICCV.2005.50 Cited on page(s) 8, 24

[89] A. Hanson and E. Riseman. VISIONS: A computer system for interpreting scenes. In *Computer Vision Systems*, 1978. Cited on page(s) 7

[90] C. Harris and M. Stephens. A combined corner and edge detector. In *Alvey Vision Conference*, 1988. Cited on page(s) 36

[91] R. I. Hartley and A. Zisserman. *Multiple View Geometry in Computer Vision*. Cambridge University Press, 2nd edition, 2004. Cited on page(s) xi, 123

[92] X. He, R. S. Zemel, and M. Á. Carreira-Perpiñán. Multiscale conditional random fields for image labeling. In *CVPR*, 2004. DOI: 10.1109/CVPR.2004.1315232 Cited on page(s) 108

[93] V. Hedau, D. Hoiem, and D. Forsyth. Recovering the spatial layout of cluttered rooms. In *ICCV*, 2009. DOI: 10.1109/ICCV.2009.5459411 Cited on page(s) xviii, xix, 8, 15, 23, 25, 28, 30, 37, 49, 51, 52, 53

[94] V. Hedau, D. Hoiem, and D. Forsyth. Thinking inside the box: Using appearance models and context based on room geometry. In *ECCV*, 2010. Cited on page(s) xxi, 23, 104, 106, 108

[95] B. Heisele, G. Kim, and A. Meyer. Object recognition with 3d models. In *BMVC*, 2009. Cited on page(s) 74, 75

[96] G. Heitz, S. Gould, A. Saxena, and D. Koller. Cascaded classification models: Combining models for holistic scene understanding. In *NIPS*, 2008. Cited on page(s) 119, 122

[97] H. Hock, L. Romanski, A. Galie, and C. Williams. Real-world schemata and scene recognition in adults and children. *Memory & Cognition*, 6:423–431, 1978. DOI: 10.3758/BF03197475 Cited on page(s) 6

[98] D. Hoiem. *Seeing the World Behind the Image: Spatial Layout for 3D Scene Understanding*. PhD thesis, Robotics Institute, Carnegie Mellon University, August 2007. Cited on page(s) xvii, xx

[99] D. Hoiem, A. A. Efros, and M. Hebert. Automatic photo pop-up. In *ACM SIGGRAPH 2005*. DOI: 10.1145/1186822.1073232 Cited on page(s) xii, xviii, 20, 23, 26, 27, 39, 42, 44, 45, 46, 117

[100] D. Hoiem, A. A. Efros, and M. Hebert. Geometric context from a single image. In *ICCV*, 2005. DOI: 10.1109/ICCV.2005.107 Cited on page(s) 29, 36

[101] D. Hoiem, A. A. Efros, and M. Hebert. Putting objects in perspective. In *CVPR*, 2006. DOI: 10.1007/s11263-008-0137-5 Cited on page(s) 25, 100, 104, 108, 111, 113, 117, 118

[102] D. Hoiem, A. A. Efros, and M. Hebert. Recovering surface layout from an image. *International Journal of Computer Vision*, 75(1):151–172, 2007. DOI: 10.1007/s11263-006-0031-y Cited on page(s) xviii, 8, 20, 37, 39, 40, 41, 42, 43, 53, 111, 116, 123

[103] D. Hoiem, A. A. Efros, and M. Hebert. Closing the loop on scene interpretation. In *CVPR*, 2008. DOI: 10.1109/CVPR.2008.4587587 Cited on page(s) xii, xxi, 20, 111, 112, 113, 114, 115, 116, 118, 122

[104] D. Hoiem, A. A. Efros, and M. Hebert. Putting objects in perspective. In *International Journal of Computer Vision*, volume 80, Oct 2008. Cited on page(s) 13, 15, 20, 22, 102, 103, 105, 108

[105] D. Hoiem, A. A. Efros, and M. Hebert. Recovering occlusion boundaries from an image. In *International Journal of Computer Vision*, volume 91, 2011. DOI: 10.1007/s11263-010-0400-4 Cited on page(s) 23

[106] D. Hoiem, C. Rother, and J. Winn. 3d layoutcrf for multi-view object class recognition and segmentation. 2007. DOI: 10.1109/CVPR.2007.383045 Cited on page(s) xx, 73, 74, 75, 79, 80, 85, 95

[107] D. Hoiem, A. N. Stein, A. A. Efros, and M. Hebert. Recovering occlusion boundaries from an image. In *ICCV*, 2007. DOI: 10.1007/s11263-010-0400-4 Cited on page(s) 8, 111, 112, 113, 114, 116

[108] A. Hollingworth and J. M. Henderson. Does consistent scene context facilitate object perception? *Journ. of Experimental Psychology: General*, 127(4):398–415, 1998. DOI: 10.1037/0096-3445.127.4.398 Cited on page(s) 6

[109] Y. Horry, K. Anjyo, and K. Arai. Tour into the picture: using a spidery mesh interface to make animation from a single image. In *ACM SIGGRAPH*, pages 225–232, 1997. DOI: 10.1145/258734.258854 Cited on page(s) 8

[110] E. Hsiao, A. Collet, and M. Hebert. Making specific features less discriminative to improve point-based 3d object recognition. In *CVPR*, 2010. Cited on page(s) 68

[111] W. Hu and S.-C. Zhu. Learning a probabilistic model mixing 3d and 2d primitives for view invariant object recognition. In *CVPR*, 2010. Cited on page(s) 74

[112] J. Huang, A. B. Lee, and D. Mumford. Statistics of range images. In *CVPR*, 2000. DOI: 10.1109/CVPR.2000.855836 Cited on page(s) 8

[113] J. Hummel and I. Biederman. Dynamic binding in a neural network for shape recognition. *Psychological Review*, 99(3):480–517, 1992. DOI: 10.1037/0033-295X.99.3.480 Cited on page(s) 58

[114] J. Hummel and B. Stankiewicz. Categorical relations in shape perception. *Spatial Vision*, 10(3):201–236, 1996. DOI: 10.1163/156856896X00141 Cited on page(s) 61

[115] D. P. Huttenlocher and S. Ullman. Object recognition using alignment. In *ICCV*, 1987. Cited on page(s) 61, 62

[116] D. Jacobs and R. Basri. 3-d to 2-d pose determination with regions. *International Journal of Computer Vision*, 34:123–145, 1999. DOI: 10.1023/A:1008135819955 Cited on page(s) 67, 95

[117] T. Kanade. Recovery of the three-dimensional shape of an object from a single view. *Artificial Intelligence*, 17:409–460, 1981. DOI: 10.1016/0004-3702(81)90031-X Cited on page(s) 6

[118] D. Kernsten, D. Knill, P. Mamassian, and I. Bulthoff. Illusory motion from shadows. *Nature*, 379:31–31, 1996. DOI: 10.1038/379031a0 Cited on page(s) 19

[119] J. Koenderink and A. V. Doorn. The internal representation of solid shape with respect to vision. *Biol. Cybern.*, 32:211–216, 1979. DOI: 10.1007/BF00337644 Cited on page(s) 60

[120] J. J. Koenderink. Pictorial relief. *Phil. Trans. of the Roy. Soc.*, pages 1071–1086, 1998. DOI: 10.1098/rsta.1998.0211 Cited on page(s) 4

[121] J. J. Koenderink, A. J. V. Doorn, and A. M. L. Kappers. Pictorial surface attitude and local depth comparisons. *Perception and Psychophysics*, 58(2):163–173, 1996. DOI: 10.3758/BF03211873 Cited on page(s) 4

[122] V. Kolmogorov. Convergent tree-reweighted message passing for energy minimization. *IEEE Trans. on Pattern Analysis and Machine Intelligence*, 28(10):1568–1583, October 2006. DOI: 10.1109/TPAMI.2006.200 Cited on page(s) 80

[123] J. Kosecka and W. Zhang. Video compass. In *ECCV*. Springer-Verlag, 2002. Cited on page(s) 14, 15

[124] P. Koutsourakis, L. Simon, O. Teboul, G. Tziritas, and N. Paragios. Single view reconstruction using shape grammars for urban environments. In *ICCV*, 2009. DOI: 10.1109/ICCV.2009.5459400 Cited on page(s) 24

[125] S. Kumar and M. Hebert. Discriminative random fields: A discriminative framework for contextual interaction in classification. In *ICCV*, 2003. Cited on page(s) 108

[126] S. Kumar and M. Hebert. A hierarchical field framework for unified context-based classification. In *ICCV*, 2005. DOI: 10.1109/ICCV.2005.9 Cited on page(s) 108, 111

[127] A. Kushal, C. Schmid, and J. Ponce. Flexible object models for category-level 3d object recognition. In *CVPR*, 2007. DOI: 10.1109/CVPR.2007.383149 Cited on page(s) xix, 65, 71, 72, 79, 80, 85

[128] J.-F. Lalonde, A. A. Efros, and S. G. Narasimhan. Estimating natural illumination from a single outdoor image. In *ICCV*, 2009. DOI: 10.1109/ICCV.2009.5459163 Cited on page(s) 19, 123

[129] J.-F. Lalonde, A. A. Efros, and S. G. Narasimhan. Detecting ground shadows in outdoor consumer photographs. In *ECCV*, 2010. DOI: 10.1007/978-3-642-15552-9_24 Cited on page(s) 123

[130] J.-F. Lalonde, D. Hoiem, A. A. Efros, C. Rother, J. Winn, and A. Criminisi. Photo clip art. In *ACM SIGGRAPH 2007*. Cited on page(s) xii, xvii, 22, 103

[131] S. Lazebnik, C. Schmid, and J. Ponce. Semi-local affine parts for object recognition. In *BMVC*, volume 2, pages 959–968, Kingston, UK, 2004. Cited on page(s) 66, 71

[132] D. C. Lee, A. Gupta, M. Hebert, and T. Kanade. Estimating spatial layout of rooms using volumetric reasoning about objects and surfaces. In *NIPS*, 2010. Cited on page(s) 20, 23, 108

[133] D. C. Lee, M. Hebert, and T. Kanade. Geometric reasoning for single image structure recovery. In *CVPR*, 2009. DOI: 10.1109/CVPRW.2009.5206872 Cited on page(s) xviii, 8, 23, 30, 37

[134] K.-C. Lee, J. Ho, and D. Kriegman. Nine points of light: Acquiring subspaces for face recognition under variable lighting. In *CVPR*, 2001. DOI: 10.1109/CVPR.2001.990518 Cited on page(s) 18

[135] Y. J. Lee and K. Grauman. Shape discovery from unlabeled image collections. In *CVPR*, 2009. DOI: 10.1109/CVPR.2009.5206698 Cited on page(s) 25

[136] B. Leibe, N. Cornelis, K. Cornelis, and L. Van Gool. Dynamic 3d scene analysis from a moving vehicle. In *CVPR*, 2007. DOI: 10.1109/CVPR.2007.383146 Cited on page(s) 99, 109

[137] B. Leibe, A. Leonardis, and B. Schiele. Combined object categorization and segmentation with an implicit shape model. In *In ECCV workshop on statistical learning in computer vision*, pages 17–32, 2004. Cited on page(s) 70, 80

[138] B. Leibe and B. Schiele. Scale invariant object categorization using a scale-adaptive mean-shift search. In *DAGM'04 Annual Pattern Recognition Symposium*, volume 3175, pages 145–153, Tuebingen, Germany, Aug 2004. Cited on page(s) 68, 75, 90

[139] V. Lepetit, P. Lagger, and P. Fua. Randomized trees for real-time keypoint recognition. In *CVPR*, pages 775–781, 2005. DOI: 10.1109/CVPR.2005.288 Cited on page(s) 80

[140] T. Leung and J. Malik. Representing and recognizing the visual appearance of materials using three-dimensional textons. *International Journal of Computer Vision*, 43(1):29–44, 2001. DOI: 10.1023/A:1011126920638 Cited on page(s) 34, 35

[141] C. Li, A. Kowdle, A. Saxena, and T. Chen. Towards holistic scene understanding: Feedback enabled cascaded classification models. In *NIPS*, 2010. Cited on page(s) xxi, 119, 121, 122

[142] L.-J. Li, R. Socher, and L. Fei-Fei. Towards total scene understanding: Classification, annotation and segmentation in an automatic framework. In *CVPR*, 2009. Cited on page(s) xxi, 99

[143] S. Z. Li. *Markov random field modeling in computer vision.* Springer-Verlag, London, UK, 1995. Cited on page(s) 28

[144] Y. Li, L. Gu, and T. Kanade. A robust shape model for multi-view car alignment. In *CVPR*, pages 2466–2473, 2009. DOI: 10.1109/CVPRW.2009.5206799 Cited on page(s) 69

[145] J. Liebelt and C. Schmid. Multi-view object class detection with a 3d geometric model. In *CVPR*, 2010. DOI: 10.1109/CVPR.2010.5539836 Cited on page(s) xx, 73, 74, 75, 79

[146] J. Liebelt, C. Schmid, and K. Schertler. Viewpoint-independent object class detection using 3d feature maps. In *CVPR*, Jun. 2008. DOI: 10.1109/CVPR.2008.4587614 Cited on page(s) xx, 73, 74

[147] J. Liebelt, C. Schmid, and K. Schertler. Viewpoint-independent object class detection using 3d feature maps. *CVPR*, June 2008. DOI: 10.1109/CVPR.2008.4587614 Cited on page(s) 85, 92

[148] X. Liu, O. Veksler, and J. Samarabandu. Graph cut with ordering constraints on labels and its applications. In *CVPR*, 2008. DOI: 10.1109/CVPR.2008.4587470 Cited on page(s) 8

[149] R. J. Lopez-Sastre, C. Redondo-Cabrera, P. Gil-Jimenez, and S. Maldonado-Bascon. ICARO: Image Collection of Annotated Real-world Objects. http://agamenon.tsc.uah.es/Personales/rlopez/data/icaro, 2010. Cited on page(s) 77

[150] D. Lowe and T. Binford. The recovery of three-dimensional structure from image curves. *IEEE Trans. on Pattern Analysis and Machine Intelligence*, 7(3):320–326, May 1985. DOI: 10.1109/TPAMI.1985.4767660 Cited on page(s) 61

[151] D. G. Lowe. Three-dimensional object recognition from single two-dimensional images. *Artificial Intelligence*, 31:355–395, 1987. DOI: 10.1016/0004-3702(87)90070-1 Cited on page(s) 62

[152] D. G. Lowe. Object recognition from local scale-invariant features. In *ICCV*, 1999. DOI: 10.1109/ICCV.1999.790410 Cited on page(s) xix, 36, 64, 65, 66, 67

[153] D. G. Lowe. Distinctive image features from scale-invariant keypoints. *International Journal of Computer Vision*, 60(2):91–110, 2004. DOI: 10.1023/B:VISI.0000029664.99615.94 Cited on page(s) 30, 35

[154] T. Lozano-Pérez and W. E. L. Grimson. Off-line planning for on-line object localization. In *FJCC*, pages 138–143, 1986. Cited on page(s) 62

[155] B. D. Lucas and T. Kanade. An iterative image registration technique with an application to stereo vision. pages 674–679, 1981. Cited on page(s) 89

[156] D. Marr. Representing visual information. In *Computer Vision Systems*, 1978. Cited on page(s) 58, 111

[157] D. Marr. *Vision*. Freeman, San Francisco, 1982. Cited on page(s) 4

[158] K. Mikolajczyk and C. Schmid. An affine invariant interest point detector. In *International Journal of Computer Vision*, pages 128–142, 2002. DOI: 10.1007/3-540-47969-4 Cited on page(s) 66

[159] K. Mikolajczyk and C. Schmid. Scale and affine invariant interest point detectors. *International Journal of Computer Vision*, 60(1):63–86, 2004. DOI: 10.1023/B:VISI.0000027790.02288.f2 Cited on page(s) 64, 65, 68

[160] P. Moreels and P. Perona. Evaluation of features detectors and descriptors based on 3d objects. In *ICCV*, 2005. DOI: 10.1109/ICCV.2005.89 Cited on page(s) 65

[161] G. Mori. Guiding model search using segmentation. In *ICCV*, 2005. DOI: 10.1109/ICCV.2005.112 Cited on page(s) 30

[162] K. Murphy, A. Torralba, and W. T. Freeman. Graphical model for recognizing scenes and objects. In *NIPS*. 2003. Cited on page(s) 8, 99, 108

[163] K. Murphy, Y. Weiss, and J. M. Loopy belief propagation for approximate inference: An empirical study. In *UAI*, 1999. Cited on page(s) 80

[164] B. Nabbe, D. Hoiem, A. A. Efros, and M. Hebert. Opportunistic use of vision to push back the path-planning horizon. In *IROS*, 2006. DOI: 10.1109/IROS.2006.281676 Cited on page(s) xii, xvii, 23

[165] S. K. Nayar, S. A. Nene, and H. Murase. Real-time 100 object recognition system. In *ICRA*, volume 3, pages 2321–2325 vol.3, 1996. DOI: 10.1109/ROBOT.1996.506510 Cited on page(s) 62

[166] V. Nedovic, A. W. M. Smeulders, A. Redert, and J. M. Geusebroek. Depth information by stage classification. In *ICCV*, 2007. DOI: 10.1109/ICCV.2007.4409056 Cited on page(s) xviii, 8, 20, 21, 25

[167] A. Y. Ng. Feature selection, L_1 vs. L_2 regularization, and rotational invariance. In *ICML*, 2004. DOI: 10.1145/1015330.1015435 Cited on page(s) 32

[168] J. Ng and S. Gong. Multi-view face detection and pose estimation using a composite support vector machine across the view sphere. In *RATFG-RTS '99: Proceedings of the International Workshop on Recognition, Analysis, and Tracking of Faces and Gestures in Real-Time Systems*, page 14, Washington, DC, USA, 1999. IEEE Computer Society. DOI: 10.1109/RATFG.1999.799218 Cited on page(s) 69

[169] D. Nistér and H. Stewénius. Scalable recognition with a vocabulary tree. In *CVPR*, 2006. DOI: 10.1109/CVPR.2006.264 Cited on page(s) 36

[170] M. Nitzberg and D. Mumford. The 2.1-D sketch. In *ICCV*. 1990. DOI: 10.1109/ICCV.1990.139511 Cited on page(s) 8

[171] E. Nowak, F. Jurie, and B. Triggs. Sampling strategies for bag-of-features image classification. In *ECCV*, 2006. DOI: 10.1007/11744085_38 Cited on page(s) 36

[172] S. Obdrzalek and Matas. Object recognition using local affine frames on distinguished regions. In *BMVC*, 2002. Cited on page(s) 66

[173] Y. Ohta. *Knowledge-Based Interpretation Of Outdoor Natural Color Scenes*. Pitman, 1985. Cited on page(s) 6, 7

[174] Y. Ohta, T. Kanade, and T. Sakai. An analysis system for scenes containing objects with substructures. In *IJCPR*, pages 752–754, 1978. Cited on page(s) xvii, 6, 7

[175] A. Oliva and A. Torralba. Modeling the shape of the scene: A holistic representation of the spatial envelope. *International Journal of Computer Vision*, 42(3):145–175, 2001. DOI: 10.1023/A:1011139631724 Cited on page(s) 8, 20, 21

[176] A. Oliva and A. Torralba. Building the gist of a scene: The role of global image features in recognition. *Progress in Brain Research*, 155, 2006. DOI: 10.1016/S0079-6123(06)55002-2 Cited on page(s) xvii, 8, 20

[177] P. Olivieri, M. Gatti, M. Straforini, and V. Torre. A method for the 3d reconstruction of indoor scenes from monocular images. In *ECCV*, pages 696–700, 1992. DOI: 10.1007/3-540-55426-2_76 Cited on page(s) 8

[178] M. Ozuysal, V. Lepetit, and P. Fua. Pose estimation for category specific multiview object localization. In *CVPR*, 2009. DOI: 10.1109/CVPRW.2009.5206633 Cited on page(s) xx, 69, 77, 78

[179] S. Palmer. Visual perception and world knowledge: notes on a model of sensory-cognitive interaction. In D. Norman and D. Rumelhart, editors, *Explorations in Cognition*, pages 279–307. LNR Res. Group, San Francisco, 1975. Cited on page(s) 58

[180] S. Palmer, E. Rosch, and P. Chase. Canonical perspective and the perception of objects. *Attention and Performance*, 9:135–151, 1981. Cited on page(s) 59

[181] S. E. Palmer. Vision science-photons to phenomenology. 1999. Cited on page(s) 79

[182] K. Pezdek, T. Whetstone, K. Reynolds, N. Askari, and T. Dougherty. Memory for real-world scenes: The role of consistency with schema expectation. *Journ. of Experimental Psychology: Learning, Memory, and Cognition*, 15(4):587–595, 1989. DOI: 10.1037/0278-7393.15.4.587 Cited on page(s) 6

[183] J. C. Platt. Probabilistic outputs for support vector machines and comparisons to regularized likelihood methods. In *Advances in Large Margin Classifiers*, pages 61–74. MIT Press, 2000. Cited on page(s) 103

[184] T. Poggio and S. Edelman. A neural network that learns to recognize three-dimensional objects. *Nature*, 343:263–266, 1990. DOI: 10.1038/343263a0 Cited on page(s) 60

[185] P. Pritchett and A. Zisserman. Wide baseline stereo matching. In *ICCV*. DOI: 10.1109/ICCV.1998.710802 Cited on page(s) 66

[186] A. Rabinovich, A. Vedaldi, C. Galleguillos, E. Wiewiora, and S. Belongie. Objects in context. In *ICCV*, 2007. Cited on page(s) 99, 108

[187] D. Ramanan. Using segmentation to verify object hypotheses. In *CVPR*, 2007. DOI: 10.1109/CVPR.2007.383271 Cited on page(s) 36

[188] L. Roberts. Machine perception of 3-D solids. In *OEOIP*, pages 159–197, 1965. Cited on page(s) 7, 24, 25

[189] C. Rother. A new approach to vanishing point detection in architectural environments. *IVC*, 20(9-10):647–655, August 2002. DOI: 10.1016/S0262-8856(02)00054-9 Cited on page(s) 15

[190] F. Rothganger, S. Lazebnik, C. Schmid, and J. Ponce. 3d object modeling and recognition using local affine-invariant image descriptors and multi-view spatial constraints. *International Journal of Computer Vision*, 66(3):231–259, March 2006. DOI: 10.1007/s11263-005-3674-1 Cited on page(s) xix, 67, 68, 81, 83

[191] C. A. Rothwell, A. Zisserman, D. A. Forsyth, J. L. Mundy, and J. L. Canonical frames for planar object recognition. 1992. Cited on page(s) 62

[192] B. C. Russell, A. Torralba, K. P. Murphy, and W. T. Freeman. LabelMe: a database and web-based tool for image annotation. Technical report, MIT, 2005. DOI: 10.1007/s11263-007-0090-8 Cited on page(s) 33, 77, 78, 113

[193] P. Y. Saad M. Khan and M. Shah. A homographic framework for the fusion of multi-view silhouettes. In *ICCV*, 2007. DOI: 10.1109/ICCV.2007.4408897 Cited on page(s) xx, 74

[194] S. Savarese and L. Fei-Fei. 3D generic object categorization, localization and pose estimation. 2007. DOI: 10.1109/ICCV.2007.4408987 Cited on page(s) xx, 69, 71, 72, 77, 78, 79, 81, 83, 85, 89, 91, 92, 93, 94, 95

[195] S. Savarese and L. Fei-Fei. View synthesis for recognizing unseen poses of object classes. In *ECCV*, 2008. DOI: 10.1007/978-3-540-88690-7_45 Cited on page(s) 71, 75, 79, 81, 83, 85, 94, 95

[196] A. Saxena, S. Chung, and A. Y. Ng. Learning depth from single monocular images. In *NIPS*, 2005. Cited on page(s) xviii, 20, 46

[197] A. Saxena, S. H. Chung, and A. Y. Ng. 3-d depth reconstruction from a single still image. *International Journal of Computer Vision*, 76, 2007. DOI: 10.1007/s11263-007-0071-y Cited on page(s) xix, 8, 21, 24, 46, 49

[198] A. Saxena, S. H. Chung, and A. Y. Ng. Make3d: Learning 3-d scene structure from a single still image. *IEEE Trans. on Pattern Analysis and Machine Intelligence*, 2008. DOI: 10.1109/TPAMI.2008.132 Cited on page(s) 28, 46, 48, 50, 53

[199] A. Saxena, J. Schulte, and A. Y. Ng. Depth estimation using monocular and stereo cues. In *IJCAI*, 2007. Cited on page(s) 20

[200] S. Saxena, M. Sun, and A. Ng. Make3d: Learning 3-d scene structure from a single still image. *IEEE Trans. on Pattern Analysis and Machine Intelligence*, 2008. DOI: 10.1109/TPAMI.2008.132 Cited on page(s) 26

[201] F. Schaffalitzky and A. Zisserman. Viewpoint invariant texture matching and wide baseline stereo. In *ICCV*. DOI: 10.1109/ICCV.2001.937686 Cited on page(s) 66

[202] C. Schmid and R. Mohr. Local grayvalue invariants for image retrieval. *IEEE Trans. on Pattern Analysis and Machine Intelligence*, 19(5):530–535, 1997. DOI: 10.1109/34.589215 Cited on page(s) 66

[203] C. Schmid, R. Mohr, and C. Bauckhage. Comparing and evaluating interest points. In *ICCV*, 1998. DOI: 10.1109/ICCV.1998.710723 Cited on page(s) 36

[204] H. Schneiderman and T. Kanade. A statistical approach to 3D object detection applied to faces and cars. In *CVPR*, pages 746–751, 2000. Cited on page(s) 69

[205] B. Schölkopf, C. J. C. Burges, and A. J. Smola, editors. *Advances in kernel methods: support vector learning*. MIT Press, Cambridge, MA, USA, 1999. Cited on page(s) 32

[206] S. Seitz and C. Dyer. View morphing. In *SIGGRAPH*, pages 21–30, 1996. Cited on page(s) 85, 87

[207] T. Shakunaga. 3-d corridor scene modeling from a single view under natural lighting conditions. *IEEE Trans. on Pattern Analysis and Machine Intelligence*, 14(2):293–298, February 1992. DOI: 10.1109/34.121796 Cited on page(s) 8

[208] L. G. Shapiro and G. C. Stockman. *Computer Vision*. Prentice Hall, 2001. Cited on page(s) xi

[209] J. Shi and J. Malik. Normalized cuts and image segmentation. *IEEE Trans. on Pattern Analysis and Machine Intelligence*, 22(8), August 2000. DOI: 10.1109/34.868688 Cited on page(s) 30

[210] I. Shimshoni and J. Ponce. Finite-resolution aspect graphs of polyhedral objects. *IEEE Trans. on Pattern Analysis and Machine Intelligence*, 19(4):315–327, 1997. DOI: 10.1109/34.588001 Cited on page(s) 62

[211] J. Shotton, J. Winn, C. Rother, and A. Criminisi. Textonboost: Joint appearance, shape and context modeling for multi-class object recognition and segmentation. In *ECCV*, 2006. DOI: 10.1007/11744023_1 Cited on page(s) 33, 36

[212] N. Snavely, S. M. Seitz, and R. Szeliski. Photo tourism: exploring photo collections in 3d. In *SIGGRAPH 2006*. DOI: 10.1145/1141911.1141964 Cited on page(s) 68

[213] A. Sorokin and D. Forsyth. Utility data annotation with amazon mechanical turk. In *First IEEE Workshop on Internet Vision at CVPR 08*, 2008. DOI: 10.1109/CVPRW.2008.4562953 Cited on page(s) 79

[214] M. Stark, M. Goesele, and B. Schiele. Back to the future: Learning shape models from 3d cad data. In *BMVC*, pages 106.1–106.11, 2010. DOI: 10.5244/C.24.106 Cited on page(s) 74

[215] F. Stein and G. Medioni. Structural hashing: Efficient three dimensional object recognition. *CVPR*, 91:244–250. DOI: 10.1109/CVPR.1991.139696 Cited on page(s) 62

[216] J. Stewman and K. Bowyer. Learning graph matching. In *ICCV*, pages 494–500, 1988. DOI: 10.1109/TPAMI.2009.28 Cited on page(s) 62

[217] P. Sturgess, K. Alahari, C. Russell, and P. H. S. Torr. What, Where, and How Many? Combining Object Detectors and CRFs, 2010. DOI: 10.1007/978-3-642-15561-1_31 Cited on page(s) 124

[218] H. Su, M. Sun, L. Fei-Fei, and S. Savarese. Learning a dense multi-view representation for detection, viewpoint classification and synthesis of object categories. In *ICCV*, 2009. DOI: 10.1109/ICCV.2009.5459168 Cited on page(s) xx, 71, 75, 79, 80, 83, 85, 91

[219] E. Sudderth, A. Torralba, W. T. Freeman, and A. Wilsky. Learning hierarchical models of scenes, objects, and parts. In *ICCV*, 2005. DOI: 10.1109/ICCV.2005.137 Cited on page(s) 8, 108

[220] E. Sudderth, A. Torralba, W. T. Freeman, and A. Wilsky. Depth from familiar objects: A hierarchical model for 3D scenes. In *CVPR*, 2006. DOI: 10.1109/CVPR.2006.97 Cited on page(s) 8, 99

[221] M. Sun, S. Y. Bao, and S. Savarese. Geometrical context feedback loop. In *BMVC*, 2010. Cited on page(s) 108

[222] M. Sun, G. Bradsky, B.-X. Xu, and S. Savarese. Depth-encoded hough voting for joint object detection and shape recovery. In *ECCV*, 2010. DOI: 10.1007/978-3-642-15555-0_48 Cited on page(s) 70, 78, 79, 80, 107

[223] M. Sun, H. Su, S. Savarese, and L. Fei-Fei. A multi-view probabilistic model for 3d object classes. In *CVPR*, 2009. DOI: 10.1109/CVPR.2009.5206723 Cited on page(s) 71, 79, 83, 85, 91, 94

[224] R. Szeliski. *Computer Vision: Algorithms and Applications*. Springer, 2010. Cited on page(s) xi

[225] M. F. Tappen, W. T. Freeman, and E. H. Adelson. Recovering intrinsic images from a single image. *IEEE Trans. on Pattern Analysis and Machine Intelligence*, 27(9):1459–1472, Sept 2005. DOI: 10.1109/TPAMI.2005.185 Cited on page(s) 18

[226] J. Tardif. Non-iterative approach for fast and accurate vanishing point detection. In *ICCV*, 2009. Cited on page(s) 15

[227] M. Tarr and S. Pinker. Mental rotation and orientation-dependence in shape recognition. *Cognitive Phycology*, 21(2):233–282, 1989. DOI: 10.1016/0010-0285(89)90009-1 Cited on page(s) 59

[228] M. Tarr and S. Pinker. When does human object recognition use a viewer-centered reference frame? *Phycological Science*, 1(4):253–256, 1990. Cited on page(s) 59

[229] D. Tell and S. Carlsson. Wide baseline point matching using affine invariants computed from intensity profiles. In *ECCV*. DOI: 10.1007/3-540-45054-8_53 Cited on page(s) 66

[230] J. Tenenbaum and H. Barrow. Experiments in interpretation guided segmentation. *Artificial Intelligence*, 8(3):241–274, June 1977. DOI: 10.1016/0004-3702(77)90031-5 Cited on page(s) 7

[231] A. Thomas, V. Ferrari, B. Leibe, T. Tuytelaars, and L. V. Gool. Using recognition to guide a robots attention. In *RSS*, 2008. Cited on page(s) xix, 65, 70, 78, 79, 80

[232] A. Thomas, V. Ferrari, B. Leibe, T. Tuytelaars, B. Schiele, and L. Van Gool. Towards multi-view object class detection. In *CVPR*, pages 1589–1596, 2006. DOI: 10.1109/CVPR.2006.311 Cited on page(s) xix, 65, 70, 71, 77, 79, 80, 85

[233] D. Thompson and J. Mundy. Three dimensional model matching from an unconstrained viewpoint. pages 208–220, 1987. DOI: 10.1109/ROBOT.1987.1088004 Cited on page(s) 62

[234] M. Toews and T. Arbel. Detection, localization, and sex classification of faces from arbitrary viewpoints and under occlusion. *IEEE Trans. on Pattern Analysis and Machine Intelligence*, 31:1567–1581, 2009. DOI: 10.1109/TPAMI.2008.233 Cited on page(s) 71

[235] R. Toldo and A. Fusiello. Robust multiple structures estimation with j-linkage. In *ECCV*, pages 537–547, 2008. DOI: 10.1007/978-3-540-88682-2_41 Cited on page(s) 81, 89

[236] C. Tomasi and T. Kanade. Detection and tracking of point features. Technical Report CMU-CS-91-132, Carnegie Mellon University, April 1991. Cited on page(s) 36

[237] P. H. S. Torr and A. Zisserman. Mlesac: A new robust estimator with application to estimating image geometry. *Computer Vision and Image Understanding*, 78:2000, 2000. DOI: 10.1006/cviu.1999.0832 Cited on page(s) 62

[238] A. Torralba, K. Murphy, and W. Freeman. Sharing features: efficient boosting procedures for multiclass object detection. In *CVPR*, 2004. DOI: 10.1109/CVPR.2004.1315241 Cited on page(s) 69

[239] A. Torralba and A. Oliva. Depth estimation from image structure. *IEEE Trans. on Pattern Analysis and Machine Intelligence*, 24(9), 2002. DOI: 10.1109/TPAMI.2002.1033214 Cited on page(s) 8, 15

[240] A. Toshev, J. Shi, and K. Daniilidis. Image matching via saliency region correspondences. In *CVPR*, 2007. DOI: 10.1109/CVPR.2007.382973 Cited on page(s) 66

[241] H. Trinh and D. A. McAllester. Unsupervised learning of stereo vision with monocular depth cues. In *BMVC*, 2009. Cited on page(s) 36

[242] I. Tsochantaridis, T. Joachims, T. Hofmann, and Y. Altun. Large margin methods for structured and interdependent output variables. *Journal of Machine Learning Research*, 6:1453–1484, 2005. Cited on page(s) 50

[243] Z. Tu, X. Chen, A. L. Yuille, and S. C. Zhu. Image parsing: Unifying segmentation, detection, and recognition. *International Journal of Computer Vision*, 63(2):113–140, 2005. DOI: 10.1007/s11263-005-6642-x Cited on page(s) 8, 108

[244] Z. Tu and S. Zhu. Image segmentation by data-driven markov chain monte carlo. *IEEE Trans. on Pattern Analysis and Machine Intelligence*, pages 657–673, May 2002. DOI: 10.1109/ICCV.2001.937614 Cited on page(s) 8

[245] T. Tuytelaars and L. Van Gool. Matching widely separated views based on affine invariant regions. *International Journal of Computer Vision*, 59(1):61–85, 2004. DOI: 10.1023/B:VISI.0000020671.28016.e8 Cited on page(s) 66

[246] S. Ullman and R. Basri. Recognition by linear combination of models. Technical report, Cambridge, MA, USA, 1989. DOI: 10.1109/34.99234 Cited on page(s) 61

[247] S. Ullman and R. Basri. Recognition by linear combinations of models. *IEEE Trans. on Pattern Analysis and Machine Intelligence*, 13(10):992–1006, 1991. DOI: 10.1109/34.99234 Cited on page(s) 60, 62

[248] C. A. Vanegas, D. G. Aliaga, and B. Benes. Building reconstruction using manhattan-world grammars. In *CVPR*, 2010. DOI: 10.1109/CVPR.2010.5540190 Cited on page(s) 24

[249] M. Varma and A. Zisserman. A statistical approach to texture classification from single images. *International Journal of Computer Vision*, 62(1-2):61–81, 2005. DOI: 10.1007/s11263-005-4635-4 Cited on page(s) 34

[250] P. Viola and M. Jones. Rapid object detection using a boosted cascade of simple features. In *CVPR*, volume 1, pages 511–518, 2001. DOI: 10.1109/CVPR.2001.990517 Cited on page(s) 69

[251] D. L. Waltz. Understanding line drawings of scenes with shadows. In P. Winston, editor, *The Psychology of Computer Vision*, pages 19–91. McGraw-Hill, New York, 1975. Cited on page(s) 19

[252] R. M. Warren and R. P. Warren. *Helmholtz on Perception: its physiology and development*. John Wiley & Sons, 1968. Cited on page(s) 3

[253] M. Weber, W. Einhaeuser, M. Welling, and P. Perona. Viewpoint-invariant learning and detection of human heads. In *Proc. 4th Int. Conf. Autom. Face and Gesture Rec.*, pages 20–27, 2000. DOI: 10.1109/AFGR.2000.840607 Cited on page(s) 69

[254] M. Weber, M. Welling, and P. Perona. Unsupervised learning of models for recognition. In *ECCV*, volume 2, pages 101–108, 2000. DOI: 10.1007/3-540-45054-8_2 Cited on page(s) 69, 84

[255] W. Wells, W. Grimson, and A. Ratan. Object detection and localization by dynamic template warping. 1998. DOI: 10.1109/CVPR.1998.698671 Cited on page(s) 75

[256] C. Wheatstone. On some remarkable, and hitherto unobserved, phenomena of binocular vision. *Philosophical Transactions of the Royal Society of London*, pages 371–394, 1838. DOI: 10.1098/rstl.1838.0019 Cited on page(s) 3

[257] J. Winn and J. Shotton. The layout consistent random field for recognizing and segmenting partially occluded objects. In *CVPR*, pages 37–44, 2006. DOI: 10.1109/CVPR.2006.305 Cited on page(s) 74

[258] P. Winston. Learning structural descriptions from examples. *The Psycology of Computer Vision*, 1975. Cited on page(s) 58

[259] H. Wolfson and Y. Lamdan. Geometric hashing: A general and efficient model-based recognition scheme. pages 238–249, 1988. DOI: 10.1109/CCV.1988.589995 Cited on page(s) 62

[260] J. Xiao, J. Chen, D.-Y. Yeung, and L. Quan. Structuring visual words in 3d for arbitrary-view object localization. In *ECCV*, 2008. DOI: 10.1007/978-3-540-88690-7_54 Cited on page(s) 74, 77, 85

[261] J. Xiao and M. Shah. Tri-view morphing. *Computer Vision and Image Understanding*, 96, 2004. DOI: 10.1016/j.cviu.2004.03.014 Cited on page(s) 87

[262] Y. Yakimovsky and J. A. Feldman. A semantics-based decision theory region analyzer. In *IJCAI*, pages 580–588, 1973. Cited on page(s) 7

[263] P. Yan, D. Khan, and M. Shah. 3d model based object class detection in an arbitrary view. *ICCV*, 2007. DOI: 10.1109/ICCV.2007.4409042 Cited on page(s) xx, 73, 74, 75, 79, 85

[264] L. Yang, P. Meer, and D. Foran. Multiple class segmentation using a unified framework over mean-shift patches. In *CVPR*, 2007. DOI: 10.1109/CVPR.2007.383229 Cited on page(s) 30

[265] Y. Yang, S. Hallman, D. Ramanan, and C. Fowlkes. Layered object detection for multi-class segmentation. In *CVPR*, 2010. DOI: 10.1109/CVPR.2010.5540070 Cited on page(s) 108

[266] B. Yao and L. Fei-Fei. Modeling mutual context of object and human pose in human-object interaction activities. In *CVPR*, 2010. DOI: 10.1109/CVPR.2010.5540235 Cited on page(s) 19, 99

[267] T. Yeh, J. Lee, and T. Darrell. Fast concurrent object localization and recognition. In *CVPR*, pages 280–287, 2009. DOI: 10.1109/CVPR.2009.5206805 Cited on page(s) 66

[268] S. Yu, H. Zhang, and J. Malik. Inferring spatial layout from a single image via depth-ordered grouping. In *CVPR Workshop on Perceptual Organization in Computer Vision*, 2008. DOI: 10.1109/CVPRW.2008.4562977 Cited on page(s) 8

[269] Z. Zhang. Floatboost learning and statistical face detection. *IEEE Trans. on Pattern Analysis and Machine Intelligence*, 26(9):1112–1123, 2004. Senior Member-Li, Stan Z. DOI: 10.1109/TPAMI.2004.68 Cited on page(s) 69

[270] J. Zhu, K. G. G. Samuel, S. Masood, and M. F. Tappen. Learning to recognize shadows in monochromatic natural images. In *CVPR*, 2010. DOI: 10.1109/CVPR.2010.5540209 Cited on page(s) 123

[271] L. L. Zhu, Y. Chen, A. Torralba, W. Freeman, and A. Yuille. Part and appearance sharing: Recursive compositional models for multi-view multi-object detection. In *CVPR*, 2010. DOI: 10.1109/CVPR.2010.5539865 Cited on page(s) 72, 77, 79

[272] G. Zimmerman, G. Legge, and P. Cavanagh. Pictorial depth cues: a new slant. *Journ. of the Optical Soc. of America A*, 12:17–26, Jan 1995. DOI: 10.1364/JOSAA.12.000017 Cited on page(s) 4

Authors' Biographies

DEREK HOIEM

Derek Hoiem is an Assistant Professor at the University of Illinois at Urbana-Champaign (UIUC). Before joining the UIUC faculty in 2009, Derek completed his Ph.D. in Robotics at Carnegie Mellon University in 2007 and was a postdoctoral fellow at the Beckman Institute from 2007–2008. Derek's work on scene understanding and object recognition was recognized with a 2006 CVPR Best Paper award, a 2008 ACM Doctoral Dissertation Award honorable mention, and a 2011 NSF CAREER award.

SILVIO SAVARESE

Silvio Savarese is an Assistant Professor of Electrical Engineering at the University of Michigan, Ann Arbor. After earning his Ph.D. in Electrical Engineering from the California Institute of Technology in 2005, he joined the University of Illinois at Urbana-Champaign from 2005–2008 as a Beckman Institute Fellow. He was a recipient of an NSF Career Award in 2011 and a Google Research Award in 2010. In 2002 he was awarded the Walker von Brimer Award for outstanding research initiative.

Printed in the United States
by Baker & Taylor Publisher Services